The FAMOUS "400"

THE TRAIN THAT SET THE PACE FOR THE WORLD
409 MILES · 390 MINUTES
CHICAGO · ST. PAUL · MINNEAPOLIS · VIA MILWAUKEE

CHICAGO & NORTH WESTERN RY.

ABOUT THE AUTHOR

As is the case with many a railroad author, George Williams' formative years were spent where the exhausts of steam locomotives and the wails of their whistles were familiar sounds. He lived less than a half mile from two huge railroad yards, in North Fond du Lac, Wisconsin, where he was born in 1919.

Williams studied engineering at Wisconsin University and planned to do design work for some locomotive manufacturer. When it became evident that steam power was on its way out, George turned to the aircraft industry. For the past eighteen years he has been an engineer with the Boeing Company.

Now living in Renton, Washington, he and his wife serve their community through their religious affiliation. Williams' hobby is model railroading and with three generations of railroading Williamses preceding him, it could be said that he came by this interest quite naturally.

LIFE ON A LOCOMOTIVE

Buddy Williams at the throttle on the 1616 Class E Pacific at the Milwaukee Depot, 1937.

LIFE ON A LOCOMOTIVE
the story of Buddy Williams,
C & NW Engineer

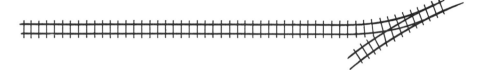

by GEORGE H. WILLIAMS

BERKELEY **Howell -North Books** CALIFORNIA

LIFE ON A LOCOMOTIVE

The Story of Buddy Williams, Engineer on the
Chicago & North Western Railway

Printed and bound in the United States of America

Library of Congress Catalog Card No. 79-165591

ISBN 0-8310-7084-6

Published by Howell-North Books

1050 Parker Street, Berkeley, California 94710

CONTENTS

PREFACE

Through the amalgamation of many railroads, the Chicago and North Western spawned a vast network of rails covering the Midwest and extending all the way to the northwestern tip of the nation. It became the fifth largest railroad system in the United States.

From its very inception, English financiers held controlling interest; consequently, the equipment and its operation have been unique in many ways. The line earned distinction in being one of the first railroads voluntarily to establish a pension system at no cost to its employees.

At a time when competitive roads were running in the red, the North Western continued to enjoy amazing financial vitality. The success was, and is, due in a large part to wise management. The employees were constantly bombarded with reminders on safety and thrift. The purchase of new locomotives would never be considered unless the existing power was either worn beyond repair or proved to be inadequate for the job. Notwithstanding, the policy of the company was to equal or better the service of its competitors, a fact which resulted in unusual demands on the engine crews.

The Lake Shore division of the Chicago and North Western cuts through the fertile farmlands and lush green pastures of Wisconsin. Dotted with beautiful lakes, the rolling landscape would seem to be an ideal country in which to railroad. The seasons were so definite that one never tired of the changing scenery; and yet the winters were so severe that only the stoutest could endure their hardships. This vigorous and interesting railroad is the background for this book, which witnesses steam at the height of its glory and sees the coming of the Diesels.

FOREWORD

Those of us who witnessed the transition from steam to Diesel power can remember how eager we were to see the new streamliners. We were awed by their graceful lines and attractive color schemes. Except for the steady rhythm of the Diesel exhaust, they appeared to glide by effortlessly.

Now we look back and realize that the passing of steam meant the passing of an era of railroading unsurpassed for its romantic aspect.

There was a fascination about the steam locomotive that defies description. Maybe it was the peculiar crack of the exhaust as the black smoke blasted the air, or possibly the echo of that lonesome whistle as it announced the train's presence to the countryside. It could have been the intriguing display of those huge drivers with the reciprocating motion of rods and valve gear. These sights and sounds gave one the impression that the engine was a powerful, living thing. Certainly these are factors that help keep the memory of steamers alive, but in the many facets of the history of our rails, one factor has been sadly neglected . . .

As the horse needed the cowboy, so did the locomotive need its engineer. Cowboy legends continue whereas the man at the steam throttle is almost forgotten.

In an effort to rectify this, I have endeavored to record my father's experiences of a lifetime in the cab and to capture the characters of his co-workers faithfully. Fortunately, George "Buddy" Williams was blessed with an animated personality, so his vivid descriptions became indelibly etched on my memory. The events recorded here are based on his reminiscences, on my own observations and on contacts with his fellow workers.

7

My father was, and is to this day, my hero. Nevertheless, he truly was considered to be second to none on a steam locomotive. In spite of the fact that he established speed records on a steam engine that stand to this day, his safety record remained untarnished. The dispatchers called him "Spark Plug."

At the time of this writing, he is in his 81st year and lives in San Manuel, Arizona. He would like to have this testimonial of esteem and admiration apply not only to him, but to all the locomotive engineers of his ilk and of his time. Their steadiness of nerve controlled the power of the locomotive and to their care the lives of the passengers and crews were entrusted.

GEORGE H. WILLIAMS

Renton, Washington
April, 1971

1.

The Prodigal Son Returns

In 1872 Levi Williams, a locomotive engineer, hired his son Charles to fire for him at Neenah, Wisconsin. Charley was only 13, but he was strong of body and mature beyond his age. This partnership continued for 3 years. Quite suddenly Levi became very ill, stricken with severe abdominal pains.

The following day there was no engineer available to take over the job. One of the railroad officials suggested that young Charley Williams could handle it if they could find someone to do the firing. A clerk in the office overhearing the conversation volunteered, "Sir, I've had experience firing those wood burners."

When Charley reported for duty the next morning, the official called him into the office and informed him that he was being considered to replace his father. "Can you run the engine?" asked the official. Charley's eyes brightened up. "Yes, sir," came his prompt reply. "My father instructed me how to operate the engine and I have had many hours at the throttle." Charley's reputation as a level-headed responsible young man influenced the final decision and he became a locomotive engineer at 16.

The job continued on without a hitch, but Charley's father grew steadily worse. One evening about two weeks after his father became sick, Charley had just returned home and as was his custom, he stopped in the woodshed for an armload of kindling. As he entered the back door, Mother Williams was

9

busy at the woodstove. "Wash up right away, dinner is almost ready," said his Mother. Charley bent over and quietly placed each stick in the bin. Lifting the lid to the reservoir, he filled the dipper with hot water and poured it into the wash basin.

"How's Father?" inquired Charley as he removed his shirt and proceeded to wash the soot from his face. Keeping her back toward her son in order to conceal her tears, she replied, "Better go on up and see him right away, he's been calling for you."

Charley finished drying and started up the stairs. "That you, son?" asked his father in a weak voice. "Turn up the lamp and sit down beside me." As Charley turned up the wick on the coal oil lamp, his father's pale face became visible. Carefully, he sat on the side of the bed.

"Charley, my boy, I'm going to die." "Now Father, you mustn't talk like that," complained Charley. "Don't interrupt, I haven't much time left." Though enfeebled, a stern note in his father's voice bespoke authority. "I have many burdens on my heart and I need your help," continued his father. "What do you want me to do?" asked Charley. "Promise me you will take care of Mother, send your three sisters through school and pay off the mortgage." Charley gave his solemn word. That night, Levi Williams passed on to his reward.

By the time Charley was 23, his mother had remarried, his sisters completed their schooling and the mortgage was paid. Having fulfilled his promise, Charley married Mary Wydotski. Ten children followed in quick succession. The fourth child from this union was named George Henry Williams.

At four years of age George imagined that the eccentric on his mother's sewing machine was the side rod of his father's engine. While pumping the foot treadle, his big toe became lodged in a hole through the platform. The up and down action nearly broke it off. It was a very painful experience, but his passion for the steam locomotive continued.

Though forbidden by his father to come near the tracks, George would often slip down to the depot and watch the engine as it glided up and down the rails. One day while the fireman was taking on water, Charley spotted son George hiding behind some bushes. Quickly he climbed down and walked over. "What are you doing here?" he said in a gruff voice. Expecting the worst, George hung his head and started to cry. Father Williams took his boy in his arms and carried him up into the cab. Standing on the fireman's seat box George looked overhead at the dangling bell cord. "Go ahead and pull it," said his father. Timidly he reached up with both hands and pulled with all his might. The huge brass bell clanged and George forgot his tears. The next hour was spent riding the engine with his father. In that short time he nearly wore the bell out.

After carrying him off the engine, his father put him down on the platform and said, "Now you get on home, right away."

From that day on, George had but one ambition in life, to be an engineer like his daddy.

George's older brother Charley turned 18 in the spring of 1905. He was his father's namesake and four years older than George. As is often the case, the big brother exercised considerable influence over him.

The railroad business was booming and the word came through that the North Western was hiring. Charley decided to hop a freight to Green Bay and hire out. When George learned of his plans, he begged to go along. Both knew full well that their father was dead set against his children hopping trains. However, Brother George assured Charley that he would keep their trip a secret, if only for his own preservation. Father Williams ruled his family with due regard to the biblical injunction "spare the rod and spoil the child."

They secretly packed a lunch and stole down to the depot. A northbound freight was sidetracked across from the depot awaiting a southbound passenger train. In order to avoid being detected, Charley led George by a route which brought them near the rear end of the train. After boosting George into an empty boxcar Charley leaped in and they both moved out of sight. When the passenger train departed the freight started out on the main line and they were on their way to Green Bay.

Charley had ridden in boxcars before but this was George's first experience. As the train cleared the Neenah yards, they sat with their feet dangling in the open doorway.

Charley figured his brother was going along for the ride, but George had made up his mind to hire out firing too. Somehow he would have to convince the man who did the hiring that he was big enough and old enough. After all, his father began firing for the North Western when he was only 13. The train made several stops and did some switching en route. It was around 6 p.m. when they pulled into a siding about a half mile beyond the Green Bay Depot. Arriving too late to apply for the job, Charley suggested they eat and then take a stroll around town.

Charley had made quite a name for himself as an amateur boxer. George had a ringside seat at all his matches, and according to him, Charley won every fight with the exception of one disputed draw. In fact, most of his opponents wound up on the flat of their backs.

Charley seemed to know where he was going, so George was content to follow. On their way to the center of town, they were passing through what appeared to be the skid row area. Charley stopped in front of a penny arcade. The place had a carnival-like atmosphere and George was eager to look things over. Right at the entrance, George spotted an intriguing machine which dispensed colored gum balls. He pulled out a penny and dropped it in the slot. Nothing happened. Charley walked over and gave the box a sharp rap on the side. This seemed to activate the mechanism and a cute little figure of a clown pivoted around and a gum ball fell into his extended hand. Returning to his former position, he dropped the gum into a trough which led to an opening in the front of the box. It was worth a penny just to see the performance. Charley stuck his coin in the slot. Nothing happened. So he gave it another rap, but no response. Then he banged it really hard with the side of his fist, but the clown just jiggled. While all this was going on, George spotted a tough looking character approaching Charley from the inside of the Arcade.

"Watch it Charley," warned George.

Charley looked up and saw the situation. With his arm down by his side, he motioned with his hand for George to get behind him.

"What do you think you are doing?" the man asked of Charley.

"Just trying to get the gum I paid for," was Charley's answer.

"Well, move along or I'll give you something you didn't pay for." As he spoke, he advanced with his arm extended as if to shove Charley backwards.

Like a flash, Charley stepped aside and unleashed a terrific right that landed on the point of the guy's chin. The man collapsed to the sidewalk, unconscious. Two more big fellows emerged from the Arcade and came directly at him. Charley moved to meet the one in the lead and caught him flush on the mouth with a haymaker. After staggering backward he went down. Seeing what had happened to his friends, the third fellow started backing up with both hands held high. "I'm neutral," he shouted. The first victim was shaking his head and trying to figure out what was going on. A small crowd was beginning to gather. Charley decided he had better clear out before some cop showed up. Taking George by the hand, he said, "Let's go, we got our money's worth."

They both slipped off down an alley. As they came out at the other end, they slowed to a walk.

"Two up and two down, that's not bad pitching," remarked George proudly.

"I would like to have retired the sides," said Charley, "but the last fellow wouldn't go to bat."

As they strolled down the main street taking in the sights, George followed along and counted himself lucky to have such a big brother.

Just before heading back to the railway yard, they passed a soda fountain. Charley ordered two ice cream cones and handed one to George.

They found an empty boxcar with a quantity of clean, loose paper, which they used to soften the hard, rough floor. George's mind was filled with the excitement of the day and the prospect of getting a job. It wasn't long before sleep overcame both of them.

Early the next morning the boxcar gave a noisy lurch and started to move. "Let's get out of here," shouted Charley as he started for the door. George followed instinctively. Charley jumped clear, but George was still half asleep and he hesitated for a moment. The train was picking up speed.

"Hurry up and jump," shouted Charley.

George leaped to the ground and the momentum caused him to stumble forward. Charley caught him just before he went down.

"Let's go," said Charley, as he trotted off in the direction of the depot. Over a flatcar, through an open boxcar and across a dozen tracks, by the time they reached the clearing, George was winded.

"Wait up for me," he yelled. Charley slowed his pace and motioned with a swing of his arm for him to hurry.

They entered the depot by the back door and headed for the men's room. After washing up, they brushed off their clothes and combed their hair. As George looked in the mirror, he realized that Charley was a head taller and had a day's growth of beard. George considered giving up the idea of hiring out. He knew his chances were slim, but he reasoned, "I have nothing to lose by trying." So he squared his shoulders and stood a little taller. Somehow that seemed to improve his hopes.

Charley asked the ticket agent about whom he should contact in order to apply for a job.

'The traveling engineer is the man you want to see," replied the agent, "he's located at the yard office about a mile down the track." As he spoke, he pointed in the direction from which they had just come.

"Let's go," said Charley. By the time they reached the yard office, George was pretty tuckered out. Charley instructed George to wait outside while he applied for the job. Usually he obeyed his big brother, and though he nodded assent, he followed quietly right behind.

A clerk opened a low swinging gate and directed Charley down the hall. George hesitated just long enough to watch Charley disappear into a room. Then he slipped in through the open door right behind him.

The official was seated with his feet crossed on the corner of his desk. The stub of a black cigar was clenched between his teeth. "What can I do for you?" he asked.

"I want a job firing," was Charley's answer.

"And what about him?" the man said, directing his question toward George. Up till now Charley was not aware that his brother was standing right behind him. Stepping forward with his shoulders back, George stood as tall as he could and said, "Sir, I want a job firing too."

"Just how old are you, son?" inquired the official.

"I became 18 the 14th of this month," was the reply.

Slowly the official removed his feet from the desk, leaned forward and said, "Tell me the truth, are you 13 or 14?"

George looked straight ahead and was about to repeat the lie when Charley interrupted, "George, you can't fool him. Now get back outside and wait like I told you."

George was crestfallen and he just stood there looking down. The official, sensing his deep disappointment, got out of his chair, walked over and said, "You come back when you are 18 and I'll give you a job."

Too embarrassed to look up, George thanked him and went on outside. After about a half an hour, Charley came out with a handful of papers and a big grin on his face.

"Did you get the job?" asked George.

"Yeah, but no thanks to you," returned Charley. "I've got a half a notion to give ya a good beating. They want me to start my student trips right away, so I'll have to get my physical exam and report right back. I'm sending you home on the next freight," he continued.

"But what will I tell Father when he asks me where I have been?"

15

Charley felt sorry for him because he would have to face the music. "Well," said Charley, "I guess the only thing you can do is go and tell Mother the truth and hope she will persuade Father to go easy on you. Now come along, I gotta get you on the train."

George followed him back to the freight yard. There was a southbound about to leave. Charley helped him into an empty boxcar and said,

"Tell mother I'll be home first chance I get. Now stay out of sight till you're out of town." With that Charley trotted off.

George's hopes for a job were gone and now he was faced with the prospect of severe discipline. After leaving Green Bay, he stood in the doorway reflecting on his misfortune. The beautiful green pastures with cattle grazing helped to ease his troubled mind. It was only 32 miles to Neenah and soon he would be home. With that thought he suddenly realized that he was hungry.

It was about noon when the train crossed the long bridge and he could see the Neenah depot. There was his father's engine parked in front of the switchmen's shanty. George figured his father was home having dinner, therefore he planned to hop off at the crossing and lay low until his dad returned for the afternoon switching. Then he would slip home unnoticed. When the train slowed down, George jumped clear and started to run. As he darted past the gatetender's shack, a familiar voice with the ring of authority sounded loud and clear.

"George! Come here!"

George froze in his tracks. As he turned he saw his father leaning out the window of the little shanty. Slowly he walked back.

"Didn't I just see you hop off a boxcar?" The tone of his voice betrayed a precariously controlled emotion.

Without looking up, he answered," Yes, Father."

"You get on home and I'll tend to you when I get off work." As he finished the sentence his arm and forefinger were pointing in the direction of home.

Charles L. Williams, father of George "Buddy" Williams is shown here during his retirement at the age of 83. He was the grandfather of the author. Below is a Chicago and North Western pass issued to him and his wife in 1926.

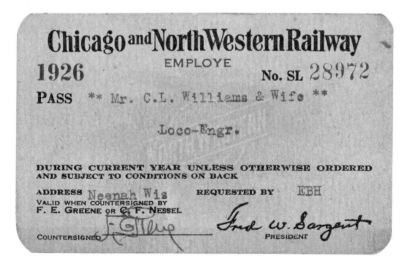

This is an R-1 type engine such as used on runs bucking snow. Pictured at Fort Atkinson, Wisconsin, it is steamed up and ready to be hooked onto the train. The pass below was issued to the author when he was attending Wisconsin University.

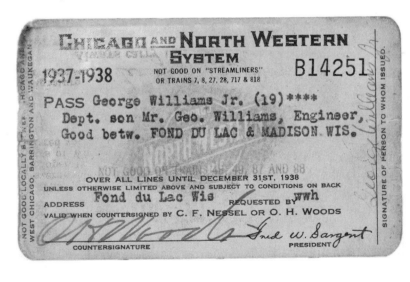

Mother Williams was in the process of finishing the noon dishes when he came through the back door. "Where have you been?" she exclaimed, almost in tears.

Painfully, he told her the whole story. "I know it was wrong," he added, "but I wanted that job so badly."

"Your father was very upset and I'm afraid you will be in for it."

As she spoke, she brushed back his thick hair and drew his head down against her breast.

"There is no getting out of it. Father caught me getting off a freight train."

Mrs. Williams was a little German woman and though she had 10 children, her concern for each was as though she had but one. "Go wash up," she said, "and I'll fix you something to eat."

While his mother stirred the wood stove and added a few sticks, George pumped some water into a basin and proceeded to clean up. When he had finished the meal George felt better. He grabbed a towel and was helping his mother when Jessie, his younger sister, came bounding through the door. After kissing her mother on the cheek, she looked at George and said, "Where have you been?"

George was in no mood to rehash the whole story, but his mother explained briefly what had been done, including the promised discipline. Jessie flinched at the thought.

"Mama, what can we do?"

"There is nothing we can do. However, I'll ask your father not to be too severe. But you know that hopping trains is a very grave offense not to mention taking off without even telling anyone."

Jessie sat down at the table with her chin propped up by her hands. For a while she seemed to be in deep concentration. Suddenly, she jumped up and said, "I've got it!"

"Got what?" asked her mother.

"I know how we can save George."

George appreciated her concern, but any hope of getting out of it seemed so remote that he felt no interest.

17

"Maybe we can't get him out of it, but we can put padding under his overalls so it won't hurt so much."

Father Williams had a heavy razor strap; one that could deliver the message through clothes quite adequately. "Padding would certainly reduce the effect," her mother added.

"Oh, Mother, you know Father would catch on and maybe then he would make me take off all my clothes."

There was a short silence. Then his mother went upstairs and came back with two pairs of brother Charley's heavy woolen pants.

"Get upstairs and put these on under your overalls and we will see how it looks."

As she finished the instructions, she tossed him the pants. George figured it was a lost cause, but he did as she said and came down the stairs waddling like a duck. When his mother first saw him she had to cover her mouth in order to hide a big grin.

Jessie spoke up, "George, Father won't even notice the difference."

"Oh, it's no use," he said, as he started back upstairs to remove the pants. "Just a minute," said Jessie, as she raced off. When she reappeared, she was carrying that mean-looking razor strap.

"Bend over, George, and we will find out just how effective that padding really is."

Jessie seemed to enjoy the prospect of laying it on her big brother. As she raised the strap and took the proper stance, George shouted his objections . . . "Just a darn minute." At this point, his mother sensed that Jessie might well convince brother George that the results may be worth the risk. "Go ahead. Bend over, George. Let's see if it will help."

The whole thing seemed so ridiculous, but at the same time, he couldn't forget how that strap had stung in the past. After a brief reflection on the subject, he said . . . "Okay, whack me just once."

WHACK! George straightened up and rubbed his rear vigorously. "Did you have to lay it on so hard?" he complained.

"Well, how about it? Did it reduce the pain?"

"Well, I guess it helped some, but after 10 or 15 of them on the same cheek it won't matter much. Maybe we better forget the whole idea."

Once again Jessie darted out and came back with one of her mother's oversized pie tins.

"Oh, no you don't," protested George.

But Jessie grabbed hold of his overalls and his mother helped jam the pie tin in the appropriate location.

"Now, let's try that again," said Jessie. The overalls were so tight that he had difficulty bending over.

Once more she got that gleam in her eye as she wound up. BLAM! George straightened up slowly with a big grin on his face. "I didn't even feel it," he said.

The whole operation had used up precious time and shortly Father Williams would be home from work. George made a final adjustment of the pie tin and sat down in the rocker to await his fate. Soon he heard familiar heavy steps.

"Where is George?" his father demanded.

"Now, Charley," said Mother Williams, as she helped her husband remove his overall jacket. Then she continued to speak in low tones. Jessie and George knew she was pleading for leniency. Father Williams pumped some water and began to wash off the coal dust. Mother Williams handed him the towel. While he was drying he said, "Now you listen to me. That boy needs a good lesson and I'm going to see that he gets it." With that he shouted, "George!"

George jumped up and answered, "Yes, Father."

"Get upstairs to my room and I'll tend to you directly."

George started for the stairs and as he reached for that first step, his father noticed the unusual contour of his lower half. The overalls were so tight that George feared each step might expose the whole plot. After reaching the room his father told him to bend over. His rear end stuck out like a bustle. Only then did Father Williams realize what had been done. Slowly he raised the strap for the first swat. As George braced himself he could see sister Jessie at the top of the stairs peeking around the corner of the banister. When nothing happened he looked around fearfully. There was his dad with his hand over his mouth. Soon Father Williams could contain himself no longer. Dropping the strap, he burst out laughing until the tears rolled down his cheeks. Slowly George straightened up with a puzzled look on his face. Never before had his father failed to carry out a threat of punishment.

Jessie ran over and planted a kiss on her father's cheek and gave him a big hug. After recovering his composure, Father Williams sat down on the bed, and said . . . "You can take off that armor plate, now. I'm not going to whip you. But don't let me ever hear of you hopping trains again. Do you understand me?" George nodded and answered,

"Yes, Father."

The outcome of George's reprieve resulted in an evening of unusual merriment in the Williams household. Maybe somewhat akin to the biblical account of *The Prodigal Son's Return.*

2.

Hiring Out

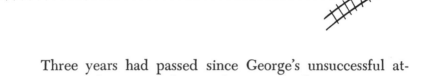

Three years had passed since George's unsuccessful attempt to go firing, and during this time he became a strong young man. The clanging of a school bell never did appeal to him. After passing his 17th birthday, the call of that lonesome whistle drew him irresistibly to his chosen work.

The year was 1908 and once again the North Western was hiring. Occasionally his father took his engine to the shops at North Fond du Lac for repairs. It was a little over 30 miles south and in order to keep from interrupting the weekly work load, these trips would be made on a Sunday. George decided to ask to go along on the next trip. While his father would be supervising the work, he planned to inquire about a job. The next time the switcher was on its way to North Fond du Lac, he was sharing the fireman's seat box.

After the engine entered the roundhouse at North Fond du Lac, George made his way over to the yard office to see Mr. Carkins, the traveling engineer. George paused to adjust his belt and brush back his hair, then, with a firm step he entered the official's office.

Mr. Carkins, a tall lean man with a soft voice and a kind face, was seated at a desk. Looking up over his horn-rimmed glasses he studied George for a moment and said, "What can I do for you?"

"I would like to have a job firing," he replied.

"You're a bit small for the job aren't you?"

George was 5 feet 7 inches and weighed around 140 pounds but it never occurred to him that his size would be a handicap.

"Oh no, sir!" he answered and with that he quickly rolled up his sleeve, bent his arm at the elbow and leaned over the desk.

"Take a hold of that and judge for yourself," he added.

Carkins pressed firmly on his biceps; a faint smile came over his face as he leaned back. George felt confident he made his point.

"What year were you born?" asked the official. The question caught George flat footed. He lost his first chance for a job because he was too young. This time he planned to claim he was 21 years old and he had actually practiced saying,

"I'm 21 years old, sir!"

But without thinking, he answered with the true date of his birth, "Born in 1890."

"Let's see now, that makes you how old?" said Carkins as he started to reach for a pencil.

"Twenty-one years old, sir!" George's answer came back so fast and with such conviction that the official put the pencil down. If Mr. Carkins detected the discrepancy, he never let on.

"You see, son," he said, "the working conditions under which a fireman must operate demand the utmost in physical stamina." And then he went on to describe the hardships and hazards involved. After a pause he looked up and asked, "Do you feel you are able to accept the responsibilities of a locomotive fireman?"

George detected that he wanted to emphasize the seriousness of this decision.

"Thank you for your concern, but I am somewhat familiar with the duties of the job and I feel confident of success." Then he added, "I want a job firing more than anything else in the world."

The official leaned back in his swivel chair and removed his glasses. Pulling a handkerchief from his hind pocket he proceeded to wipe off the lenses. George feared he was stalling to formulate a way of letting him down easy. The suspense was agonizing.

Then suddenly Carkins scooted his chair back from the desk, reached in his drawer, pulled out a form and shoved it in front of George. "Here, fill this out."

The questionnaire was brief and he completed it within a few minutes. After reading the application, Mr. Carkins looked up and said, "Aren't you Charley Williams' son, our engineer at Neenah?"

"Yes, Sir!" replied George.

"Why, I made my student trips firing for your father," commented Mr. Carkins. After a pause during which he studied George with renewed interest, the official grabbed the phone and called the company physician.

"Hello, Doc Pullian? This is Carkins. I'm sending Charley Williams' boy down for an examination so we can give him a job. Can you handle him this morning?"

"Good, I'll send him right down."

With that he took a form out of his desk, signed it and handed it to George. "Take this to Doctor Pullian and get back here as soon as you can."

George darted off. Everything went like clockwork and he was back with the Doc's okay before noon. Carkins met him at the door, glanced at the medical report and said, "Follow me."

They walked into an adjacent office where a man sat busy at his desk. "Wallie, I want you to meet Charley Williams' boy, George. I have just put him on firing. George, this is Walter Hoffman, our division foreman."

"Put 'er there," said Mr. Hoffman as he stuck out his hand. "If you're anything like your dad, the North Western will be fortunate to have you."

"By the way, Herb, who has been firing for Charley?"

"That new fellow named Butski," replied Carkins.

"Ain't it about time for him to get some experience on the main line?" inquired Hoffman. As he finished the sentence he gave Carkins a sly wink.

Carkins developed a broad grin and answered, "That's just what I was thinking."

Fireman Butski was having dinner at the McGivern Hotel. Carkins picked up Hoffman's phone and called him.

"Hello, Butski? this is Carkins. I have decided you should get out on the main line and learn the road. So, I'm relieving you from the Neenah switcher and marking you up on the extra board."

As he put down the phone, he turned to George and said, "I'm assigning you the Neenah switcher to fire for your father. Now, they are just about finished with his engine, so, I suggest you get out there and go to work."

Things seemed to happen so fast that George was a bit overwhelmed, but the words, "GO TO WORK" snapped him out of it. Both men shook his hand and George started for the roundhouse.

As he arrived back at the engine, one of the mechanics shouted, "She's all ready, Charley."

Climbing up into the cab, George noticed the steam was back to 90 pounds, so he spread a layer of coal over the remaining fire.

While his father was inspecting the work that was done, George skipped over to get a fresh jug of water. Upon returning he saw heavy black smoke rising from the stack. A couple of twists on the blower valve cleared it up. When his father returned to the cab, he looked around and said,

"Where's my fireman Butski?"

George was busy cleaning the water gage and pretended not to hear him.

"George!"

"Yes, Dad."

"Have you seen my fireman?"

At this point, Charley Williams knew nothing about his son hiring out.

Moving over to the engineer's side, George directed an extended thumb toward his chest, and replied in an authoritative tone, "I'm your new fireman. What are you waiting for?"

Charley was one of those Englishmen who rarely ever showed any emotion. His father responded by reaching in his overall jacket for a package of chewing tobacco. Meticulously he prepared a good size quid and crammed it in his mouth. After two or three chews, he looked down at his son and said, "Well, have you got her ready?"

George detected a twinkle in his eye as he spoke. "I've got 'er all set Dad," he answered.

With that George went back to his seat box and checked on his side to be sure all personnel were clear.

"All clear!" he shouted, as he gave the bell cord a yank.

The little switcher moved out onto the turntable and lined up for the track in front of the coal shed. George was starting over the coal gate by the time the tank was spotted under the chute.

"Be sure to stand clear before you dump that coal," shouted his father.

George reached up and gave the handle a jerk and the coal came rumbling down with a cloud of dust. When the tender became full, he jumped back down on the deck and proceeded to clean up the chunks that had spilled over the gate. Charley noticed George's face was more black than white.

"Better wash your face before we get home or Mother won't recognize you."

The engine moved up to the penstock and George climbed back up on the tank to take water. When he returned, his father was studying the time card.

"We can meet 216 at Tower DX and check there with the dispatcher for a lineup," explained his father. "Let's get going," he added as he opened the throttle.

It was about three-fourths of a mile uphill to Tower DX. The little switch engine moved out gracefully and covered the distance in short time.

Personnel operating with a switch engine are in extreme danger. On two separate occasions when Charley Williams laid off, a man in the crew was killed, but during his 56 years on a switch engine, no one ever lost his life while he was at the throttle.

As the engine came to a stop directly across from the tower, Charley moved the reverse lever to high center and started to get off. Facing the gangway in order to climb down, he hesitated long enough to give George some instructions. "Keep a watch on the engine while I get a lineup from the dispatcher. In the meantime, check your water level and get your fire in shape. I want to be ready to pull out when 216 gets by."

Firing a switch engine was not entirely new to George. Frequently his father had let his fireman take off early on Saturday afternoon so that he could catch 216 to Fond du Lac. On these occasions, George would have the opportunity to fire the last two or three hours of that day.

George prepared his fire carefully and put on the injector. While waiting for his father's return he grabbed the squirt hose and washed down the deck.

When Charley climbed back up into the cab he announced, "The railroad is all ours after 216 passes."

A whistle sounded in the distance. Looking ahead, George could see 216 rocking along toward them at a good clip.

Lon Sage, the engineer on 216, fired for Charley many times and they were good friends. When the passenger approached, Charley gave him a couple of toots. The ground shook as the train sped on by. When the tail end of the last car passed, the fading sound of an answering toot-toot could be heard.

Charley moved the engine to the north end of the siding. George ran ahead and threw the switch. When the engine passed he closed it again and climbed back on. Soon they were rocking along about 35 mph and George settled down to the routine of firing.

Running light, that is without a train, the little engine was easy on steam and didn't require much attention. While enjoying the scenery, George was mentally figuring up what his first paycheck would be.

The swing bridge over the channel at South Oshkosh was aligned. As they slowed down for the city, George got busy on the bell cord. His father gave a low whistle for each crossing. Leaving Oshkosh he widened on the throttle. They moved right along. Soon they were coming into Neenah. As they crossed the main intersection of town, George hoped that he might be seen by some of his chums, but instead the driver of a delivery wagon with a tired looking nag was the only spectator.

Easing past the depot they stopped for the lead which headed for the roundhouse. Again, George took care of the switch.

The engine came to a halt over the ashpit and his father helped him knock the fire.* After his father backed the engine into the roundhouse, George opened the blower valve in order to exhaust the remaining steam out the stack. Charley shut off the air pumps and closed the valves on the feed water lubricators. While this was being done, George climbed off and closed the huge doors to the roundhouse.

On their way home, his father outlined some of the things he would need.

"You'll hafta have some heavy duty shoes, and gauntlet gloves. Mother can get those for you. First day we get off in time, we'll go down and pick out a good 21-jewel watch."

George respected his father and he was anxious to be a worthy son. He had hoped to see some external sign of approval, but Charley was a very austere individual and though he was proud of George he purposely concealed it. As they entered the house through the back door, his mother was busy at the wood stove and George could smell onions frying.

"Mary!" said his father, "our boy got a job firing today."

Her face almost glowed with joy. Wiping her hands on her apron, she extended her arms toward her son. As they embraced, George lifted her off the floor and made a complete turn.

---

*Shake the grates and dump the ash pan.

"This calls for a celebration, Charley," said his mother excitedly.

Charley pulled out his pocketbook, handed Jessie a half a dollar and said, "Run down to the store and get a couple quarts of cream."

George brought the freezer up from the basement. After chipping off a good size chunk from the icebox, he proceeded to chop it and spread it around the container. Meanwhile Mother Williams got busy and stirred up a batch of mix. By the time Jessie returned, George had sprinkled rock salt on the ice and was ready to crank the handle. Ice cream was a rarity in the Williams' household.

George was in the spotlight for the remainder of the evening and he enjoyed every bit of it.

3.

The Challenge

By five o'clock in the morning, Mother Williams would have breakfast cooking. George slept directly above the master bedroom where there was an open register between the floors. When the coffee started boiling, George's mother would awaken her husband and say, "Charley, the coffee is on."

George was a light sleeper and her voice would alert him also. There would be just enough light to enable George to gather his clothes and carry them downstairs. The Williams' home had a central heating unit. That is, it was located in the middle of the living room. A huge pot-bellied stove radiated heat in all directions. The closer one came, the warmer, and vice versa. Right beside this heater, a big easy chair provided George with an ideal spot to finish dressing. Washing up was accomplished by getting a pan of water from the reservoir in the cooking stove and pouring it in a basin. Washbowls, running water and inside toilets were luxuries reserved for the wealthy.

Breakfast usually consisted of a big bowl of oatmeal, toast and coffee. The toasting was done on a long-handled fork. Mother Williams would remove one of the stove lids and hold the bread over the coals. Fresh ground coffee, boiled in a porcelain pot, helped wash down the dry toast.

The regular working day was from 6 a.m. to 6 p.m. With just a five-minute walk, they would be at the roundhouse. Usually, the helper would have the fire pretty well started. George would take over from there and by the time the steam reached 40 or 50 pounds, his father would start the air pumps and begin oiling the running gear with the long-spouted oil can.

During that time, George had several duties to perform. First, he would put in a good layer of coal and turn up the blower. Next he would remove the jug from the locker on the tank and go over to the pump for fresh water. Passing through the storeroom on the way back, he would jam a handful of clean waste* in his pocket. The waste was used to clean the faces of the gauges and the waterglass. Generally he would finish up by wiping off all the cab windows. By that time, his father would put the oil can back in the rack and check on the air pressure. When he was ready, he would say, "How's she look on your side?"

Charley would never move the engine until he was sure everyone was in the clear. When George gave the "all clear," the little engine would move out of the roundhouse over to the switchman's shanty.

While the conductor was outlining the day's work with Charley, George would check on his supply of coal and water.

Two switch engines were required to handle the business at Neenah. Both crews reported to work at the same time. When the briefing was completed, George would have the steam right on the 180-pound marker. His father would shout, "Let's go to work."

The crew responded by hopping on the footboards of the tank and the day's work started. It wasn't long before George knew the area like the palm of his hand.

One morning about a month after George started to work, his father signalled to line the switch for the passing track just across from the depot. There were four or five boxcars ahead of the engine. No. 206, the express job, was due shortly. When the switch engine cleared the main, Charley set the brake and instructed George to watch the engine while he went to the restroom in the depot.

---

*Discarded yarn from a stocking factory.

30

George moved over to the engineer's seat box, propped his feet up on the boiler head and leaned back. No. 206 pulled in, loaded the express and in about five minutes was highballing out of town. Unaware that the engineer was in the depot, Schaefer, the brakeman gave a signal to come ahead. George looked over toward the station, but his father was nowhere in sight. The brakeman continued giving signals and showing signs of getting impatient. George decided to accommodate him. So he released the air, moved the reverse lever forward and opened the throttle. Simultaneously, the other switcher began pushing three or four gondolas alongside of him on the main. Schaefer started giving George what appeared to be kick signals. This called for acceleration. The two engines were almost side by side going at the same speed. Hackbush, the engineer on the other switcher gave a toot-toot, which George took to be a challenge. So he yanked her wide open.

The little engine snorted like a bull and started pulling in the lead. But he hadn't gone very far when the brakeman gave him a big washout signal and leaped off the flat car. The last George saw of him, he was rolling down in the ditch. The track ahead curved to the left and George's vision was obscured by a large building. Slamming the brake valve in emergency, George braced his foot against the boiler head. The brakes just started to take hold when the head car smashed into a string of empties. The impact nearly sent him through the forward window. The noise could be heard for blocks. Coal came tumbling over the coal gate.

Hackbush stopped his engine and the crew came over to check on the switchman who took the high dive into the weeds.

As it turned out, no one was hurt. George went over to inspect the damage. Schaefer approached with a noticeable limp and said. "Why in the world didn't you slow down?"

"Because," George replied, "you were giving me kick signals, and in my book that calls for acceleration."

"Kick signals! My God, man, those were car length signals."

Kick signals are given by the arm held out and moving the hand vertically up and down rapidly. Whereas car length signals uses a similar movement, but with a much slower motion. Each cycle of the arm indicates the number of car lengths left before contact.

George pointed out that his arm waving was too rapid to be understood as car length lengths. A young switchman in the other crew named Donnahue was waiting for a chance to get in his dig, and he remarked, "You hammerhead, you don't know one signal from another."

George's temper flared and he started toward Donnahue. Suddenly the booming voice of his father interrupted his progress.

"What's going on here?" Father Williams was just leaving the depot when he heard the crash. Turning to the conductor he added, "What happened?" They both walked over to inspect the damage. George and Donnahue stood there eyeing each other.

"Looks as if we'll have to replace a couple of draw bars. Outside of that, it doesn't look too bad," commented Charley.

"I'll back down to the roundhouse and pick up the car repairman. I'm sure he can take care of this."

Grabbing hold of George's arm, he turned him in the direction of the cab and shouted, "Let's go back to work. The show is over."

The car repairman was a capable fellow and he managed to repair the damage in short order. Charley saw to it that no report was ever made of the incident.

Not a word passed between George and his father until they were on their way home after work. Father Williams spoke up, "What ever caused you to pull that foolish stunt?"

George was anxious to explain and he re-enacted the signals which he had misunderstood. They continued walking for nearly a block in complete silence. His father was a man of few words, but when it came to discipline, his vocabulary was adequate and the emphasis was sufficient. George braced himself for a well-deserved tongue lashing. Guilt and shame weighed heavily on his shoulders.

"I promise, it won't happen again," he said.

Charley sensed his son's true repentance and decided not to belabor the subject further.

"I know what you mean, son. Just last week I had to warn Schaefer on his sloppy hand signals."

George was beginning to feel relieved.

"By the way," added his father casually, "who won the race?"

Up till now nothing was said about a race.

"What race are you talking about?" asked George with a well-feigned innocent expression.

"Now, George," said his father, "you never fooled me for a minute. I heard Fred give you the challenge with the whistle and I knew from the sound of those two exhausts that you were both wide open."

George was afraid to answer. Finally, just before they entered the back door his father repeated the question, "Well, who did win the race?"

"I was taking him good until we banged into those gondolas," he replied apologetically. Just for a second George thought he detected a muffled chuckle.

That night, before going to sleep, there was one burden that still bothered George. That loudmouth brakeman wouldn't settle for less than a good beating and it looked like fate had assigned the job to him.

4.

Taming a Tough Brakeman

George continued firing for his father until his job was bulletined* and another man bid it in. This meant that he would now be working off the extra board.

The extra board usually consisted of seven or eight men who were made available for new assignments such as on extras, doubleheaders, work trains, or possibly as replacements for men off sick.

Each time an extra man completed his cycle of duty, he would report in and the clerk would mark him up on the board. Then he would await his turn for his next job. This process is called bucking the extra board, and it is the route that all new men must take.

The regular assignments were awarded on the basis of seniority, consequently the older heads had the pick of the more desirable jobs.

Good fortune smiled on George and he caught the other switcher working opposite his father in the Neenah yard. Eating and sleeping at home was a big help financially, especially at this time. His watch set him back $75, and new work clothes added another $25.

Fred Hackbush was his new engineer. This fellow's consuming passion in life was baseball and everything else was incidental. He operated the engine with just one thought in

---

*Listed on the bulletin board in order to alert those who might be interested in the available assignment, to be awarded on the basis of seniority.

mind — namely, get the job done. Unless the subject was baseball, his conversation was limited to the bare essentials. His fellow workmen jokingly referred to him as the two-position throttle man. Meaning he was either wide open or shut off.

Firing for Fred was a new experience. Knowing Hackbush's reputation, George was prepared for the worst. By this time, he had enough savvy to compensate for Hackbush's peculiar throttle technique. After the first day on the new job, George appreciated the smooth way his father operated his engine.

The North Western was rebuilding the long wooden bridge into a steel structure. Bridge builders and engineering crews were busy all over the north end of the yard. Frequently the switcher would be called on to move equipment in support of the work.

George made friends easily and soon he was acquainted with the whole gang. Except for Donnahue's hostile attitude, his crew was a congenial lot. One evening, walking home from work, George brought up the subject of Donnahue's attitude. Father Williams was aware of the situation and he advised George to stay out of his way. Not that he lacked confidence in his son's ability to protect himself, rather, he was anxious that George keep his record with the railroad company clean.

"Just do as I tell you and give that bully a wide berth."

George realized that this was wise counsel and he determined to heed his father's advice.

It always seemed to me that a brakeman or a switchman was low man on the totem pole. His pay check was the smallest of the crew. He was the one who was always climbing on top of the cars and jumping off and onto moving trains. Often he would have to run ahead of the engine, throw the switch and stand there until it passed. He was the one who always had to work in rain, sleet, or snow. Enginemen would often chide the switchmen about being the conductor's messenger boys. In spite of this, a spirit of good-natured kidding would usually exist between enginemen and train crew.

Younger brakemen were often more sensitive and one could hardly blame them, if, on occasion, they growled a little. They couldn't smart off to the conductor because he was their boss. The engineer was a poor target because he had too many whiskers.* That left the fireman. This condition is the position George found himself in with this tough Irish brakeman. Whenever this fellow rode the cab his comments would contain some form of criticism toward enginemen in general and firemen in particular. Nearly every sentence would be punctuated with a foul oath. After about two weeks on the job, Hackbush made one of his few comments:

"You know, George, that fellow Donnahue has always been disagreeable. But since you came, he's getting downright nasty." Looking over at George he continued, "I'd like to see someone teach him a good lesson."

"The truth is," returned George, "I have discussed this matter with my father, and he advised me against tangling with him. But I've had just about all I can take. The next time he shoots off his mouth, I intend to administer that lesson you mentioned."

"He's a real toughy, but I believe you can do the job," said Fred.

It was a crisp early morning in December and a light snow was falling. They had been switching material into position for the new bridge. George enjoyed watching the progress of the construction. It was a sizable undertaking with about fifty men on the job. On this occasion they were backing down the main line north of the depot, when the brakeman signaled to spot the gangway between the cab and the tank, directly across from the section crew's tool box. It was being used as a coal bin for the gate tender's shanty. There was a crosswind blowing and George had the windows closed and the curtain pulled across the back of the cab.

Perched comfortably on the seat box, George was watching the crane hoist a huge beam into place. Donnahue shouted

---

*Had seniority

36

something from below the cab. George failed to get the message. So he opened his side window and inquired as to what it was he wanted. Donnahue bellered back so everyone could hear:

"Get off your lazy hind end and throw down some coal for the gate tender and be D — — —quick about it."

George leaned out the window and said politely, "Would you mind repeating that?"

Donnahue's face turned red and he shouted louder than before. This time he added, "If that ain't plain enough, just step off that engine and I'll bang it into your thick skull."

By this time the commotion had attracted considerable attention. With a calm, almost disinterested voice, George countered, "You know, I might just let you have a little coal, but you will have to come up here and throw it down yourself."

Donnahue went into a rage, tore off his overcoat and proceeded to call George everything but nice. When he hesitated for want of more abusive language, George pointed his finger at him and said casually, "You have insulted my mother and for that you will have to apologize." George's coolness was due to the fact that he knew the showdown was coming and he was ready. He had been quite successful as an amateur boxer. With brother Charley as his tutor, he had trained on the punching bag until he had arms and shoulders way out of proportion to his size.

As George moved off the seat box, Hackbush met him on the deck and said, "Do the job right and if anyone tries to interfere, I'll take care of him."

Railroad men are instructed to climb off an engine facing the step. Before his feet had hit the ground, Donnahue stepped up behind him and tried to land one. But George saw it coming and jumped back so that the fist glanced off his shoulder.

Donnahue regained his balance and lunged for him with a haymaker. It was a clumsy move that gave George ample warning. He stepped aside gingerly and caught Donnahue on the back of his head. The blow sent him sprawling into the cinders. George backed off and let him get to his feet. This time Donnahue decided to move in slowly and just start slugging. But George kept him at a distance by pumping his left hand right on his nose. Donnahue started swinging wildly. His nose began to bleed profusely. George decided to cross over with a right to his head. As he threw the punch, his foot slipped and he stumbled to one knee. Before he could get up Donnahue landed a roundhouse right, high on George's forehead. It rocked him backward on his heels. Donnahue tried to crowd him and follow up his advantage. George back-pedalled and gave him another bang on the nose for his effort. Donnahue backed up, blew his nose and wiped the blood off his face.

Charley Hall, the conductor went over and patted Donnahue on the shoulder saying, "Go back and finish him off." Schaefer, the other switchman shouted, "You got him going, Donnahue, don't let up on him."

George noticed Donnahue's mouth was wide open and he was breathing hard. Apparently the bloody nose was cutting off his air.

Suddenly it occurred to conductor Hall that he was blocking the main line and the express was due any minute. He signalled for Hackbush to back into the siding. Fred jumped over to his side and spun those little drivers all the way into the clear. After setting the brake he jumped down to witness the final round. They were both sparring at a distance, when 206 began whistling for the crossing. As the train approached, George stepped back on one side of the track and Donnahue moved to the opposite side. While the coaches were gliding by, Hackbush came over to George and said, "Don't pay any attention to that flat-foot Hall. Donnahue couldn't hit the inside of a barn with a handful of rice."

"Oh, yeah? Well, tell me Fred, who put this big knot on my head?" As George spoke, he pointed to a reddened swelling on his forehead.

"Just a lucky punch," returned Hackbush.

When the rear end of the train passed, Donnahue stepped up into the middle of the track with his fists doubled. George moved cautiously forward to a place about six feet from him.

"Go after him," shouted Hall. The encouragement seemed to inspire Donnahue and he charged at George with both arms swinging. This time when George stepped aside he put all he had into a right cross and it connected over Donnahue's ear. The force of the blow knocked him down on the ties. Again George stepped back. Donnahue stumbled to his feet, grabbed George in a headlock and started wrestling. They both tripped over the rail and rolled down into the empty coal box. George landed on top and Donnahue couldn't move.

"What do you want to do, wrestle or box?" said Donnahue.

"You name it," answered George.

Donnahue knew as a wrestler he was done for.

"Okay, let me up and we will finish this with our fists."

Hall and Schaefer both helped Donnahue to his feet. There must have been over a hundred spectators by this time. Fortunately for George, his father was busy switching at the south end of the yard. This time when George squared off with Donnahue, he decided to go on the offensive. So instead of waiting for him to move, George waded into him. The change in tactics caught Donnahue off guard and George landed a hard right to his jaw. Donnahue fought back gamely but that blow stunned him and before he could recover, George rained lefts and rights to his head. Finally he just sagged to one knee with his head down. He was cut over his eye and blood from his nose nearly covered the front of his shirt. George waited for him to get up. The crowd became silent as they sensed the end was near. Finally Donnahue looked up with a pained expression. "I've had enough," he said slowly. "Please accept my apology." As he spoke he put out his hand. He looked so pitiful that George felt ashamed.

"All right, Jack. I accept your apology," said George, and they shook hands. *

Hackbush was doing a slow burn as he watched Hall put his coat over Donnahue and help him to his feet.

---

*In the years that followed, this tough Irishman became a highly respected conductor who was liked by all who knew him.

39

"Hall, you dirty . . . ! You encouraged Donnahue until he got a good beating. Now I am going to give you the same thing Williams gave him. With that Hackbush started for the conductor. Hall wanted no part of Hackbush. So he took off on the dead run right across the long bridge. Hall was hopping from tie to tie with the lanky Hackbush in hot pursuit. Somehow the conductor managed to stay ahead. Hackbush finally gave it up and returned to the engine.

George was busy getting his fire back in shape. It was over a week before Hall dared to climb up into the cab.

5.

Firing on the Main Line

It was early in the month of January, right after one of those fierce Wisconsin snow storms. George had been firing for Hackbush over two months. On this particular morning, the temperature was around 10 below zero and the wind made the bitter cold all the more severe.

They had just coupled into five or six loads of paper and the brakeman headed them in on the passing track across from the Neenah depot. After stopping the engine, Hackbush glanced at his watch and said, "Tell the boys we will hold here for 206."

George opened his window and conveyed the message to the switchman and they made a beeline to get thawed out in the depot. Working out in this weather was miserable, but as long as the cab windows were shut and the back canvas curtain was closed, the heat from the boiler made it almost tolerable.

With nothing to do but wait, George propped his feet up on the boiler head, leaned back and began to reflect on his financial progress. His watch, a gold-cased 21 jewel Elgin, cost $75, and it was paid for. The very thought prompted him to pull it out, wind it and run his thumb over the face of the crystal.* After putting his watch back, he took out his wallet

---

*Over a half century after the events of this chapter transpired, Dad handed his watch to me and said, "Here, son, she's yours. Take care of it and don't forget that many's the time I put my life right on that second hand.

41

and counted the cash. The total came to $48. He had been paying his mom $5 a week for room and board. Considering a day's pay was only $3.71, George figured he has doing all right. But it was about time he got a crack at the main line. Those jobs paid better and were more exciting.

No. 206 should have been there by this time and Hackbush was getting impatient.

"Run over and get us a line-up on 206," said Hackbush. George buttoned his jacket, put on his gloves, pulled down on the bill of his cap, and made his way in snow almost knee deep. As he came through the door, the ticket agent was busy loading coal in the big pot-bellied stove.

"What's the story on 206?" requested George.

"Hold on," replied the agent as he closed the door on the stove, "I'll call the dispatcher."

After handing George the report, the agent commented, "Looks like Spooner is having trouble again today." Frank Spooner was the engineer on 206 and he had a reputation for being on time.

George glanced at the report:

"No. 206 left Appleton 10 minutes ago."

Back on the engine, after reading the report, Hackbush commented, "We might as well wait them out."

With that he pulled out his jug, poured a cup of coffee and handed it to George. Fred always carried an extra cup just for his fireman. George opened the firebox door and stood with his back to the fire while he sipped the coffee.

"I used to fire for Spooner," said Fred, "and I can tell ya, when he's late, he has no mercy on his fireman."

George thought to himself, "If Spooner was any rougher on firemen than Fred he would have to go some."

In a few moments 206 could be heard whistling for the north end crossing. As the engine rounded the curve on the long bridge, George could see the front end was packed with snow nearly up to the headlight.

"Looks like he's been bucking pretty bad snow," commented George. "They should double head 206 on days like this," returned Hackbush.

The brake shoes set up a chatter as the train came to a quick stop. George checked his watch; they were over half an hour late. There weren't many passengers, but the mail and express usually took around ten minutes.

George just finished his coffee when he heard someone outside shout, "Hey, Williams!" Pulling back the curtain, George saw Bill Schaefer holding up a slip of paper. After stooping down for the message, he moved back into the cab and read it aloud.

"Fireman on 206 is unable to continue past Neenah, due to sickness. Request Fireman George Williams on Neenah switcher to provide emergency relief. A substitute fireman will arrive on 151 to take over the duties vacated by Williams." It was signed by Herb Carkins, the traveling engineer.

The first thing that came into George's mind was the fact that 206 could not depart until he took over the firing responsibilities. Working on the main line was just what he had been hoping for. But George had figured on some nice easy-going way freight. Now he was required to fire on the fastest express job on the whole division and on an engine in whose cab he had never set foot. Somehow George didn't feel too happy about his new prospects. Suddenly the words came back to him: "When Spooner's late he shows no mercy." The thought nearly threw him into panic.

"Lord, help me," he murmured under his breath. Turning to Hackbush he said, "Do you think I can keep Spooner in steam?"

"You can handle it all right, George. Only let me warn you. Bank your fires heavy in the back corners and light in the forward end. Spooner has a heavy hand on the throttle and he tends to let those drivers break loose on the start. The draft will try to pull your fire up the stack. Also keep your water level at a half a water glass or lower. Those Atlantics* have dirty boilers and they foam easily. Better get going now. They're waiting for ya."

---

* 4-4-2

Working away from home brought up many problems for which George was not prepared. "What job do I come back on?" he inquired.

"You'll have a 7-hour layover at Milwaukee and come back on 209. That will get you back here around 7:00 tonight," answered Hackbush.

As George started down the steps, Fred added, "Good luck."

Just before he stepped off in the snow, George hollered back, "Fred, would you please ask my father to put a change of clothes, razor, towel and soap in a grip and have it down here at the depot so I can pick it up on my way through tonight?"

"I'll take care of that," answered Hackbush.

The head end of the passenger was six car lengths down the track. When George climbed up into the cab, the sick fireman was gone. Spooner, a powerfully-built man with a large nose, sat facing forward with his arms folded on his chest. Jutting out of the corner of his mouth was a deep-bowl corncob pipe.

"I'm your relief fireman," announced George.

Spooner turned to face him and slowly removed his pipe. After scrutinizing him from head to foot and back again, he exclaimed, "Here I am nearly an hour late and they send me a boy to do a man's job." After a pause he added sarcastically, "Well, get on the business end of that scoop and let me know as soon as you got 'er ready."

The situation didn't allow any time for George to feel sorry for himself. Turning his attention to the firebox door, he studied the difference. Instead of a chain hanging from the end of the throttle quadrant to the door handle, this engine had a foot treadle which controlled two butterfly type doors. He stepped on the treadle and the door snapped open. The fire was pretty well burned down. Looking up at the steam gage, he noted the working pressure was 200 pounds but the needle was back to 150.

George started his scoop in for a full load, then hesitated. Looking up at Spooner, he said, "I'll need about ten minutes more."

The loaded shovel was heavier, the firebox was bigger, everything was strange. After putting in a good layer and building up the back corners, he opened the blower valve. The water in the boiler was barely visible above the bottom of the water glass. Instead of being alongside the boiler waist high, the injector was down alongside his seat box. After putting on the injector, he inspected his fire and placed a few scoops in a low spot. By this time the water had climbed almost halfway up the water glass. Recalling Hackbush's warning, he quickly turned off the injector. Then he jumped back on the deck and heaved in another round of coal. The fire was burning evenly and the pointer on the steam gage was nudging the 200-pound mark. Satisfied that he had made the proper recovery he turned to the engineer and said, "I'm ready."

Spooner slid his window open and gave the conductor a wave of his hand. Turning back, he slammed the window shut and gave two quick yanks on the whistle cord. Glancing through his back window, George could see the little switcher outlined against the snow. A thin wisp of steam was drifting skyward from the leaky pop valves. Somehow he had developed an affection for that little engine and the sudden change left him with mixed emotions.

As the brakes were released, a blast of air alerted George to the fact that he was starting on his first mainline experience. Spooner opened the throttle but the engine stalled after taking the slack in the first two or three cars. Without shutting off, he threw the reverse lever (Johnson bar) in the backward motion and backed until he had the slack in on four or five cars. Then with a grunt he heaved the bar over in the forward corner and yanked the throttle half open.

The engine responded sluggishly and threatened to stall after each exhaust. Finally she began to accelerate. They hadn't travelled 100 feet when Spooner jerked the throttle wide open. Almost instantly the drivers lost traction and started spinning. Before he managed to shut off, huge clouds of black smoke were billowing high and George feared that his fire might have gone up with it. The big 81-inch drivers finally settled down and soon Spooner had the throttle back out to the last notch.

George made a quick inspection. Some of the coals were pulled forward but most of the fire was still intact. Hurriedly, he set about to repair the damage.

They had travelled less than a mile and were making around 40 miles per hour, when George noticed the steam was falling back at an alarming rate. Something had to be done quickly. George opened the firebox door, and used the shovel as a baffle, in order to deflect the air rushing in. (This little trick would reduce the terrific glare of a roaring fire.) What he saw was most discouraging. Instead of building up his fire, the engine was consuming the coal so fast that he could see bare spots on the grates.

George began to bale coal with all his strength. He no sooner spread a good even layer when Spooner yelled, "Hey, Buddy," and pointed to the water glass. This indicator can be read at a glance while the engine is standing still, but on the move the water surges in the boiler and one must study the glass for five or ten seconds before he can come up with the approximate water level. At this point George could see the water was almost out of sight.

The roar of the exhaust makes conversation difficult and enginemen often rely on hand signals to convey their messages. George held up his right hand and twisted his wrist to indicate he understood.

After cutting in his injector, it took two or three minutes before the water level even started to rise. Occasionally they would hit a drift and fine snow would filter in around the back curtain.

Due to the tall drivers, these Atlantics had a high center of gravity. The deck of the cab was about six feet above the rail. Any irregularity in the track would cause a pronounced rolling action in the cab. The engine headed into a tight curve just as George was swinging with a full scoop. The cab lunged sideways and he rammed the shovel into the side of the opening. Coal went flying all over the deck. While scraping up the mess he noticed the leading edge of the shovel was badly bent from the force of the blow.

George tried to figure why he was getting such poor results from all the coal he was burning. Surely, he thought to himself, the other firemen were not expected to heave coal from Green Bay to Milwaukee without a chance to rest. How in the world did they stand up to it?

They would be rolling nearly 50 mph before Spooner would position the reverse lever within three or four notches of dead center. Anytime he was working over half throttle and the Johnson bar was over seven or eight notches forward, it was impossible to maintain the proper working pressure.

Glancing over the coal gate, George could see that the coal was not shifting forward as it should and soon it would be beyond his reach. His only chance to take care of the problem would be during the stop. While they were easing into the Oshkosh depot, he made his way to the top of the pile and started moving the coal forward. A large chunk shifted under his foot and he lost his balance. The action started a slide which carried him head first down against the coal gate. Half covered with coal, still clutching his shovel, George wiggled his way clear. Except for a few bruises and a covering of coal dust, he was none the worse. Strange enough, the coal ended up neatly stacked against the gate, just the way he wanted it.

When he climbed back into the cab, Spooner was busy knocking his pipe against his heel. Apparently he was not aware of the little episode back on the tank. George rammed his shovel into the pile to start stoking, but it refused to slide under the coal. Then he remembered that the flange was bent up. Quickly he grabbed the monkey wrench and by using the scuff plate in the opening of the firebox as an anvil, he flattened the damaged portion. Sighting along the edge he could see it was far from straight, but for now it would have to do.

Once again he dug the scoop in and started stoking. After the first layer was evenly spread, he turned the blower way up. Then going over to the gangway he checked on the progress of the mail and express. He had four or five minutes at the most. The gage showed 190 pounds but the water was low. On with the injector and back with the scoop. Picking up the steam was always slower with that cold water rushing into the boiler. Just as the pop valves let go, Spooner hollered, "Highball!"

The train seemed to start a little easier and the pops settled down. They were over two miles out of town, going nearly 60 mph, before George could shut off the injector. The exhaust at this speed was almost a constant roar. The crossings were coming so fast that Spooner just rested his hand on the whistle cord between each whistling post. It was only 18 miles to Fond du Lac, but at the rate this muzzle loader was eating coal, he would need a rest by then.

Tower DX flashed by and they started downhill into Fond du Lac. As the speed picked up, Spooner eased off to half throttle. George grabbed a moment's rest on the seat box. Passing the yards at North Fond du Lac, he returned a wave to a friendly engineer on the switch engine. From this point on, the road was entirely new to George. Spooner started whistling for the Scott street crossing and made his first brake application.

"Hey, Buddy," he shouted, and motioned for George to come over to his side. "We'll take on water here. Now be careful on the back of this tank. It is covered with ice."

George continued his short rest as the train drifted the last couple of miles into the station. The engineer spotted the tank under the penstock. George jumped down, grabbed the long handle and pulled the spout over. The top of the tank was just as he said. After making his way carefully, he flipped open the lid, hauled the spout down into the opening and pulled the handle to start the water flowing.

At this height he could see Spooner making the rounds with the oil can. Looking back, he could see the baggage and express men busy with their chores. The tank filled sooner than he expected and the overflow gushed up before he could shut it off. The wind whipped the water into a spray that seemed to engulf him, freezing almost instantly. His overalls felt like a suit of armor and he was chilled to the bone.

Back in the cab, he anchored the firebox doors open and thawed out in front of the fire. Spooner pushed the curtain aside and put the oil can away.

"Better get a good load in her 'cause it's all uphill out of here for nearly eight miles." There was a ring of frost around Spooner's mouth and nostrils, and the bowl of his pipe showed a dull glow with each puff.

This firebox was twice as big as the one on the switcher and in order to keep a level fire, the placement of the coal was much more critical. George was beginning to adopt a new approach. In order to reach the far end, it was necessary to swing the shovel harder, but this tended to land the whole contents in a pile. However, if he carefully judged the elevation of the scoop as it penetrated the opening, the heel of the shovel would hit the plate and the resulting bump would actually lift the coal upward. This action would also spread it out more. The idea of trying to keep this engine hot going uphill for eight miles gave him a sinking feeling. After preparing the fire with an extra heavy load, he opened the blower valve and put on the injector. Spooner studied his watch for a moment and then shouted, "The conductor is on your side; watch for his signal. We should be getting out of here any time now."

Looking back, George could see that they were nearly finished loading. Spooner sat with his left hand on the brake valve and his pipe clenched between his teeth.

"Those lazy louts had enough time to load a freight train!" growled Spooner.

The pop valves started to sizzle. George manually opened the firebox door and set the handle to hold that position. Hopefully, the cold air rushing in would hold off the pops until the engineer opened the throttle.

The conductor gave the highball and George repeated it aloud. Spooner released the brake and grabbed the throttle. As the drivers took hold, the train began to move slowly.

George could see the hill starting less than one thousand feet ahead.

Spooner was working the engine harder than ever. They barely managed to hold 30 mph on the climb. As they neared the top, the steam had dropped back in spite of everything George could do. When the engine finally passed the crest of the hill, the pressure was back 30 pounds. The engineer had to add his injector just to hold the proper water level. It must have burned a ton of coal during the climb. Spooner hooked up the Johnson bar four or five notches and the exhaust became sharper as they picked up speed.

The needle on the steam gage started to climb back where it belonged. George almost collapsed as he plopped down on the seat box. If that hill had been one mile longer, he would have lost all his steam. Hackbush was right, thought George, they should doublehead under conditions like this. He began to wonder if he might not become a casualty like the fireman he relieved at Neenah. Dropping back 30 pounds is a serious loss of power. George was a bit shaken up and the glamour of firing on the Main Line was fast losing its appeal.

There was a brief stop at Campbellsport and another at Kewaskum. Spooner never let up on the throttle. George stood on that pitching deck and heaved coal until he was ringing wet with sweat. His face burned from the constant exposure to that roaring inferno. Spooner sat there rocking from side to side with his left hand resting on the whistle cord and his eyes glued to the track ahead. They started down a long hill and as they picked up speed, the engine began swaying badly. Several times he had to check his swing in order to keep from missing the firebox door. George shut off the injector and as he leaned over the seat box looking ahead, he could see the outskirts of Milwaukee. About this time Spooner eased off to half throttle. George was exhausted and he took the opportunity to rest on the seat box. He had been so busy, there was no time to appreciate the sensation of speed. But that hi-wheeler must have been rolling close to 80 mph. The swaying was so bad that he felt he could reach out and touch the telegraph poles on either side. As they approached the one-mile-to-station marker, Spooner grabbed off about 15 pounds of air and lapped the brake valve. Soon the train was slowing down. There were six or seven sets of tracks leading right in front of the main passenger gate at the Milwaukee station. The North Western had a huge arched shelter covering the passenger loading area, and there was a constant shuffle of baggage and express wagons. The whole operation was a source of constant danger to personnel. The train was down to around 10 mph as they entered under the shed and George was busy on the bell. Spooner stopped just as the engine emerged from the south end of the shelter. After studying his watch he remarked, "Not too bad. We made up five minutes between here and Neenah."

George managed a weak smile. He thought to himself, "You big toad. I hope you didn't over exert yourself pulling the whistle cord."

Spooner took a little slack,* the brakeman pulled the pin on the back of the tank and they moved ahead on an adjacent track. Then he realigned the switch so that the relief engine could back in and couple up to the train. The approach to the roundhouse crossed over a vast network of tracks. Spooner stopped the engine under the penstock directly in front of the turntable. After stowing the engine Spooner pulled his grip out of the seat box and started to get off. Glancing over the coal gate, he remarked, "In another 25 miles we would have been out of fuel." As George followed him off he was tempted to add, "You would have been out of a fireman before that coal was gone."

Spooner lead the way over to the enginemen's washroom. The sun had come out and the wind had died down, but George had been sweating and he was anxious to get in where it was warm.

Spooner moved with a slow deliberate gait, carefully picking his way over rails that were hidden by snow. The black-

---

*Backed the engine enough to relieve the tension on the couplers between the tank and first car.

ened brick roundhouse seemed to engulf them. Soon Spooner approached a door and jammed his shoulder against it.

Good thing it swings inward, thought George, or he would have busted it off its hinges. The washroom was a dingy place with small half-blackened windows near the top of a high ceiling. After tossing his grip on a nearby bench, Spooner noticed George had brought nothing with him.

"By golly, we stole you off that switch engine there at Neenah and never gave you a chance to pack your grip."

"I'll manage," replied George, as he unhooked the straps on his overalls."

"Well," continued Spooner, "if you don't mind sharing the same soap and towel, we'll both get washed up for dinner."

As he spoke, George got a glimpse of his face in a mirror. Everything but his eyes was black. After working up a good lather and applying it generously, Spooner handed George the soap and said, "Do the best you can, Buddy. I have already forgotten what you look like." George wasn't sure if he was trying to be humorous or just plain candid.

After drying off, Spooner handed him the towel.

"By the way, Buddy, my name is Frank Spooner."

"I'm George Williams," he answered, and they shook hands.

It was getting close to noon and George had worked up a huge appetite. "You got any suggestions where a fellow might get a good meal at a reasonable price?" he asked.

"Nick The Greek has a good steak house just a block beyond the depot," replied Spooner. "You must be a stranger to this end of the road," he added.

"This is my first trip on the Main Line," returned George.

Spooner put his grip in a locker and said, "Follow me."

As they approached the shed, a northbound passenger stood ready for departure. The station master could be heard announcing the cities along the route.

Cold snow crunched under their feet as they crossed the intersection and made their way down the street to Nick's Steak House where the aroma of fresh brewed coffee met them just inside the door. Spooner picked out a stool toward the far end of the room, and George sat beside him. It was too early for the noon crowd and they had the counter to themselves.

"What will you have, Boys?" asked the waitress as she slid two steaming cups of coffee before them.

"Make mine T-bone medium with French fries," answered Spooner.

"I'll take the same," added George. "What time do we report for duty?"

"If 209 gets in here on time, they call us for 4:45," answered Spooner.

"Do we get the same engine?"

"Yep," answered Spooner. "Going or coming, we get the same one."

A change of engines might have helped George determine why he was having difficulty keeping the steam up.

The doors to the kitchen flung open and the waitress barged through with their dinners. George tackled that steak like he hadn't eaten for days. After trimming the bone bare, he looked over and saw Spooner finishing his off between both hands. As Spooner started loading his pipe, George reached for a toothpick and commented, "That was a good dinner; now all I need is a little rest."

"You can get a cot at the depot for 25 cents. It's clean and if you notify the callboy where you'll be, he'll wake you when it's time," said Spooner. "I gotta tend to some business downtown; see ya at the roundhouse," he added.

"I'll be there," replied George.

On the way back to the depot, George was thinking about the return trip. The warmth of the depot was most welcome. After contacting the callboy, George located the room with the cots. Before undressing, he carefully placed his watch and wallet under the pillow. With his hands clasped behind his head, he relaxed, staring up at the ceiling.

"Herb Carkins was right," thought George. "Firing certainly does demand the utmost in physical stamina." His first experience with firing on the Main Line had shaken his confidence.

53

6.

No Mercy

The callboy shook George and said, "You're called for 4:45; 209 is on time."

After rubbing his eyes, George looked up at him with a puzzled expression.

"You're firing on 209 for Engineer Spooner, aren't you?" said the callboy. The name of Spooner was the key which brought back the events of the day.

"I'm the victim," confessed George. Then he thanked him and started to dress. Pulling on dirty overalls made him most uncomfortable. However, he consoled himself with the hope that there would be a change of clothes waiting for him at Neenah.

George made his way through the crowded waiting room to a lunch counter at the far end of the depot. A hot cup of coffee helped clear away the remaining cobwebs. With a toothpick sticking out the corner of his mouth,* he started for the roundhouse. When he reported for duty, the clerk informed him that his engine was ready on the lead track.

At the turn of the century, passenger trains were being pulled by the famous high-wheeled Atlantics. These engines were designed to handle four or five wooden coaches, but the trains were getting longer and heavier steel coaches were replacing wooden cars. Schedules were beginning to slip and breakdowns were frequent. The North Western solved this problem by purchasing a promising new locomotive known as

*A toothpick became George's trademark.

the Pacific. Up till now, only a few of these engines had been delivered. On his way through the roundhouse, George was hoping to check one out. The place was a beehive of activity. Some engines were blowing down,* others were being overhauled. One had the smokebox opened exposing the exhaust nozzle and flues. Just a few stalls beyond that, he spotted a shiny new Pacific. She had six-foot drivers. Her lines were graceful, much like a race horse, and yet it was apparent that she was much stronger than the Atlantics. After walking around the engine, George climbed up to inspect the cab. Taking the engineer's position on the seat box, George sighted along the boiler.

"Well, what do you think of her?" Leaning out the cab window, George looked down and saw Spooner standing with grip in hand.

"She looks strong to me, sure wish we could have her going back tonight," replied George.

As he started to get off, his eye caught some raised letters cast in the iron tray over the firebox door, which read: "ROUND HOLE GRATES, KEEP FIRE LIGHT."

"Good advice," he reasoned, "as long as the hoghead doesn't slip the drivers and send the fire out the stack."

"We have about 25 minutes to get our engine ready," commented Spooner. With that, they made their way over to the lead track. Before climbing up into the cab, George noticed the coal was piled high on the tender.

"How is the coal out of here?" he asked.

"Not as good as the stuff we get out of Green Bay," replied Spooner, as he placed his grip in the seat box. "Too much ash and it tends to form clinkers."**

George went back on the tank to check the water. When he returned Spooner was starting down with the oil can. After making a quick check, George noticed that the boiler was half

---

*Exhausting steam.

**Clinkers rob the fire of precious burning area and if not taken care of, they will soon get out of hand.

full and there was a thin fire spread over the grates. Grabbing the scoop, he started to shove it in under the coal pile, but the force required reminded him that the flange was still crooked. Looking out the gangway, he could see the rear end of a tank just inside the roundhouse. Pulling the damaged scoop free, he jumped down with it and slipped inside the roundhouse. Finding the other engine deserted, he quickly exchanged shovels and hurried back. Any guilt about his thievery was soon dismissed.

"After all," thought George, "the guy who gets that shovel won't be firing for Spooner and for that he should be thankful."

Again he took care to bank his fire heavy in the back corners. After turning on the blower, he hosed down the deck and proceeded to clean the faces of all the gages.

Spooner put the oil can in the rack, checked his watch, and said, "We better get started."

As he climbed up on his seat box, he shouted, "Are you ready?"

Leaning out the window, George checked his side and shouted back, "All set!"

The sun was starting to hide behind the horizon and as they moved on down the track, the long shadow of the engine formed a picturesque silhouette on the snow. Spooner stopped with the blow-off cocks lined up with the ducted shield.

"Let's give 'er a good shot here," said Spooner as he reached forward and pulled open the blow-off cock. George reached down and followed suit.

With almost 200 pounds of pressure pushing the water out of the bottom of the boiler, the accumulated impurities would be forced out. This action tends to reduce the foaming of the boiling water and the result is that the cylinders are fed dryer steam. Dryer steam gives better expansion, hence more power.

The roar of the twin jets filled the whole yard and the escaping steam produced an impressive display. When Spooner shut off, the water was almost out of sight in the gage. George put on his injector and added another layer of coal. Spooner stopped the engine on the north end of the No. 3 main where

209 would soon pull in. The schedule going back was just as fast, and they usually had a heavier train. Up till now, Spooner expressed no dissatisfaction with his performance. But the return trip was bound to be tough. George knew that somehow he would have to do better.

No. 209 came drifting in slowly and stopped about 200 feet back. Soon the engine uncoupled and moved ahead on the adjacent track. The brakeman gave George a back-up signal and he relayed it to Spooner. After coupling into the train, Spooner checked his air gages.

"Come over here, Buddy," he said. George figured he was about to get a good bawling out. While standing patiently at his side, Spooner pulled out his pouch and started packing his pipe. Slowly, he put the pipe in his mouth and set his jaw. Then after brushing off some loose tobacco he leaned over and looked down into George's face.

"Now, I want you to get this straight," said Spooner. "Every night the officials in the general office ask the dispatcher, 'what time did 209 reach Green Bay?' and when the answer is 'late' they get very unhappy. "We have a heavy train," he continued, "and a poor excuse for an engine. But I intend to pull into Green Bay on time if I hafta beat this iron horse to death doing it."

George nodded meekly and went back to his seat box. As the last few passengers were getting on board, he tried to analyze the real meaning of Spooner's message.

Finally, he concluded, "What that old hoghead really has in mind is that me and the engine will both die if he doesn't make it in on time."

The situation would be comical except for the fact that the engineer was Lord and Master, and if Spooner decided that his fireman wasn't cutting the mustard, this would be his last trip.

The conductor came up to the engine with the orders. Spooner climbed down and read them aloud. After comparing watches, he climbed back and handed them to George. While Spooner was getting his tobacco lit, George was watching for the highball. The acrid fumes from the corncob began to fill the cab. George opened the window for a breath of air. About that time, the conductor gave the highball and George shouted, "Let's go!"

Spooner backed the engine until he had the slack in all but the last two or three cars. Then with a grunt he heaved that Johnson bar over in the forward motion. Due to his abundant girth, it was necessary for him to release the lever before it reached the far corner. But the momentum carried it to the end of the quadrant with a thud. Then with both hands he cracked the throttle. The engine responded slowly but Spooner managed to jockey the throttle enough to keep from stalling. Each time the wheels would threaten to spin, Spooner would let out with "Settle down, you slippery . . . !"

They were going over 20 mph before he began hooking her up.

This expression refers to adjusting the Johnson bar back toward the center of the quadrant. The valves of a steam locomotive have an adjustable stroke, which is controlled by the reverse lever, sometimes called the Johnson bar. When the position of the lever is in the extreme end, the valve has the largest opening. This admits the maximum amount of steam to the cylinders. However, as the speed increases, the exhausting of the steam cannot keep pace with the intake — therefore, to maintain maximum power, the engineer must shorten the stroke of the valve in order to minimize back pressure. At this stage of development, the steam locomotive had no gages by which an engineer could synchronize the throttle and reverse lever. The key to proper operation was determined by evaluating the the kind of sound coming out of the stack. When the speed increased, there would be a corresponding change in the crack of the exhaust. Maximum efficiency was achieved through a delicate balance between the position of the reverse lever and the throttle. These adjustments should be altered with each significant change in speed. Some engineers operated as though they were deaf. Others found difficulty in translating the sounds into the proper action. However, there were a select few who trained themselves so as to obtain the last ounce of available power.

It was upgrade leaving Milwaukee and in spite of all George could do, the steam dropped back 15 pounds. A couple of times Spooner had to add his injector to keep a proper water level. It was getting dark and George would snatch an occasional view of the track ahead. The oil-burning headlight was hardly any good beyond 200 feet. When the train hit level track, they began picking up speed, and Spooner had the reverse lever four or five notches ahead of center with the throttle wide open. Slowly the needle started moving back toward the 200-pound marker.

With the firebox door open, the reflection from the canvas curtain would provide just enough light to show the grim expression on Spooner's face. He just sat there, left hand on the whistle cord, puffing that pipe and rocking from side to side. George would breathe a sigh of relief each time he would shut off for a station.

They were about 15 miles out of Fond du Lac when a large clinker started forming in the center of the firebox. George carefully guided each shovelful around the spot, but the clinker continued to grow. As they started down Eden Hill, Spooner hooked her back near the center and eased off to about half throttle. George took the opportunity to work on the clinker.

All coal-burning engines carried a clinker bar back on the tank. This was a round iron rod about one-half inch in diameter, about ten feet long, with two fingers forming a claw-like end.

Thrusting the bar into the firebox, George tried to break up that fused mass of ash. Glaring heat burned his face and his cotton gloves started to smoke. The engine had gained considerable speed and the deck was pitching so badly that he nearly fell down several times.

"We'll take water here at Fond du Lac," yelled Spooner. George nodded and kept hacking away at the clinker. The engine went into a hard curve and the cab lurched sideways, sending George sprawling over against the engineer's seat box.

Spooner never took his eye off the rail. George struggled back to his feet. The clinker was loosened, but the fall convinced him that he might better be able to finish the job with the engine standing still.

When they stopped at the station George jumped down and pulled the water spout over. Then after climbing up the back of the tank, he opened the lid and hauled down the spout.

Ice seemed to cover everything. The handle for releasing the water was frozen closed, so George stood on his toes and gave it a yank. Water came with a gush and again the spray engulfed him. It was dark and the only way he could be sure the tank was full was by letting it run over. This time he managed to shut off before getting drenched. Swinging the spout clear, he climbed back into the cab.

Time was short and George was getting desperate. After anchoring the firebox door open, he placed the huge shaker bar on the shaft of the grate and shook it carefully. The clinker was shifting and he kept working the bar till it finally fell into several chunks. Then he grabbed the clinker bar and pounded them into smaller pieces. With a few short movements of the grates, most of them fell into the ash pan. Quickly, George stowed the bar back on the tank and started stoking his fire.

Spooner put the oil can back in the rack, grabbed a piece of waste and wiped off his gloves.

"Keep a lookout for the highball," he said, as he pulled his pipe out of his pocket.

George's face burned, his eyes were smarting and he began to wonder if he could last another 65 miles. The fire started shaping up and George tossed in another round. The pops started to sizzle. A twist on the injector quieted them down.

When the crew was through loading, the conductor gave a long swing of his lantern. "Highball!" yelled George as he pulled the bell cord. The main line cut through the industrial section of Fond du Lac. As the engine strained to gain speed, the echo of the exhaust bounced back from the adjacent buildings.

While passing through the freight yard at North Fond du Lac, the yellow beam of the headlight picked up what appeared to be an engine crew heading home after a day's work. By this time they were making around 50 mph. The steam went back about 10 pounds leaving town. Starting uphill toward Tower DX, the engine began to labor. The slightest climb was enough to slow her down. It was about two miles to the top and George was doing everything he could to keep the steam up. As they began to lose speed, Spooner advanced the reverse lever. Again the clinker began forming and George was helpless to do anything about it. They were down to around 35 mph and Spooner was about to give her a few more notches forward when he leaned over and saw the steam back to 170 pounds. George watched Spooner's hands as they relaxed their hold on the lever. Could it be that this tough ole hoghead was going to show a little mercy?

George caught a glimpse of Tower DX as they went by. This meant they were over the hill. Again the coal was getting hard to reach. Each swing made with that scoop was accompanied with pain from his tired back. Slowly, ever so slowly, the pointer on the steam gauge started back.

Spooner hooked her up a couple of times and the exhaust began to sharpen up. Soon they would be approaching the swing bridge at South Oshkosh. George slumped down on the seat box. In an instant he relaxed so completely that the swaying of the cab nearly rolled him off onto the deck. This was one of the few opportunities he had to see above the firebox door. A bright moon was lighting up beautiful snow-covered Lake Winnebago. It was just a fleeting view but it helped to remind him that life consisted of something else beside being a slave to a coal-eating monster.

By the time they reached the swing bridge, Spooner slowed down the train to around 30 mph. Crossing the steel structure, the clicking of the rails was mingled with the creaking of each joint on the bridge. Soon the lights of the platform at Oshkosh could be seen. George took advantage of the fact that Spooner drifted the last two miles. Before they made a complete stop, he had the clinker bar back in the firebox and had started to pull, push and pry. After struggling three or four minutes he tossed the smoking rod back on the tank and grabbed the shaker bar. This time he shook it harder. The clinker broke into several smaller pieces. Spooner never left the engine, just sat there fussing with his pipe.

This was a short stop. George got busy and put in a good load. Spooner moved over to the gangway on the fireman's side and watched for the signal.

The conductor gave the highball and Spooner shouted "Let's go!"

The fire was ready; the steam was right on the 200-pound mark, and he had a half a glass (water glass indicator) of water. Just maybe, George thought to himself, I'll manage to keep him in steam this one time. Spooner released the air, George put on the injector and started the bell. Spooner got on the throttle too hard and the drivers broke loose as if they were on grease. Heavy black smoke billowed high and formed a cloud big enough to hide the moon. There was no doubt in George's mind that part of his well-prepared fire went out the stack. The only question was how much was left. Finally Spooner got the engine to settle down. At this point George became so discouraged that he had to force himself to get off that seat box and examine the damage. It was as he had feared, nearly half of the coals went up the stack. The only thing that didn't shift forward was the remains of the clinker. George worked feverishly to build it back up. The steam dropped back 35 pounds. In desperation, George shut off the injector. Spooner started hooking her up sooner than usual. The needle hung around 165 pounds for several miles. George straightened up momentarily; his back felt as if it were broken. Looking up, he saw Spooner silhouetted against the sky, rocking from side to side with that pipe sticking straight out.

"How could he be so nonchalant in the midst of all this chaos?" For just a few seconds, George entertained the idea of handing him the scoop at Neenah, and telling him, "Now I know why your last fireman had to give it up, cause I have the same kind of sickness." But he couldn't bear the thought of the shame he would bring upon his father. So again, he mustered his remaining strength and continued the struggle. The lights of the station at State Hospital flashed by, only seven miles to Neenah. The fire was beginning to shape up.

George found time for a short breather. Looking back, the train was trailing around a gentle curve. The gas lights of the coaches projected the outline of each window on the snow and the lamp on the last coach could be seen gliding around the remainder of the curve. Looking forward, he could recognize a few of the landmarks in the bright moonlight. Spooner began to ease off. Suddenly George remembered that his only hope of clean clothes depended on whether or not Hackbush got the message to his dad. Another trip without a change would be more than he could stand. As Spooner set the brake, George spotted a few of the gas lights which lined the main thorough-fare of Neenah.

Crossing the intersection by the Valley Inn, one could see the marquee of Neenah's only theater. Going north, the Neenah depot comes in view after the train negotiates a long curve to the left.

Just before they eased to a stop, George detected someone running alongside the tank with a suitcase and a lunch box. The engine drifted on down a couple of car lengths so as not to block the crossing. When George got on the ground, there was his younger brother Ralph.

"Hi, George, how did it go?" Ralph almost idolized his big brother, and George was reluctant to tell him anything discouraging.

"It's a little rougher than firing that hobby horse for Hack-bush," he answered. "What have you got there?"

"Mother packed some clean clothes — a wash cloth and towel."

"What have you got in the lunch box?" asked George.

"There's a couple of pork chops and I don't know what all."

"Good boy, Ralph. Tell Mom thanks. I got to run now," said George as he grabbed the grip and heaved it up on the apron of the tank. With the lunch box under his arm, George started up the steps.

Ralph just stood there with his hands in his pockets. Before disappearing behind the curtain, George looked back and shouted, "Better get on home now before you catch a cold."

George put his grip in the seat box and set the lunch box on the floor alongside the boiler. Spooner was oiling the running gear. George barely had time to prepare his fire. The conductor met Spooner just as he was about to climb up into the cab.

"We're all ready Frank," he said.

George was putting a couple of extra scoops in the back corners. Spooner gave two short blasts of the whistle and started the train. As they were pulling away, someone shouted, "Hey, George!"

George opened his window. There was Ralph running alongside of the cab. "Mom said to eat the pork chops right away, while they are still warm," shouted Ralph.

"All right, I will," answered George, "now get on home."

Just like mother to see to it that her boy had a warm lunch, thought George. The train was still crossing the long bridge when he pulled out a pork chop and started in on it. In a matter of seconds he had it cleaned to the bone. With mouth still half full, he tossed in another round, then took another couple of minutes to finish off the other chop and have a swallow of hot coffee. As he closed the lid on his lunch box, he said to himself, "That will have to last me till we reach Green Bay." Spooner was working the engine hard and as usual the steam pressure was dropping fast.

The little nourishment tasted great, but now he had to make up the time he had taken for it. When the steam pressure was where it belonged, the pointer would be straight up, registering 200 pounds. But as the presure began to drop, the pointer moved counterclockwise. Whenever he would try to steal a moment's rest that pesky needle started aiming right at him. George had developed a mental picture of the inner workings of the gage. There was a devilish little imp in there and he was exerting all his strength against the wheel which moved the arrow toward him. All the while, he was singing gleefully, "Gonna get ya, gonna get ya!" The tempo seemed to coincide with the exhaust of the engine. George was getting very weary. Another short stop at Kaukauna and on the start Spooner let the big drivers break loose again. One glance in the firebox told him the bad news. It looked hopeless but he started to build it back up. The little imp seemed to be singing a second verse, "Now I got ya, now I got ya!" In the midst of his stoking, George had a real urge to take his shovel and bash that gage against the boiler head.

Spooner was a product of the old school as engineers go, and to him "on time" meant just what it said. However, contrary to the reputation he carried, he was not entirely without a heart. Though he concealed his concern, he was aware of the struggle his fireman was having.

As they drifted into the Green Bay depot, the steam was back to 160 pounds. After uncoupling, they started for the roundhouse. Not a word was exchanged as they washed up. George realized the job was more than he could handle. The idea of giving up was revolting. But he just couldn't go through another trip like that. As he finished washing up, he made up his mind to quit railroading. Somehow, his father would have to understand. Spooner was busy making out his engine report. Too discouraged to even think of the future, George leaned back on the bench and waited for Spooner to finish.

After folding the paper, Spooner handed it to the clerk and said, "Wire this report to the road foreman of engines at once!"

Picking up his grip, he turned to George and said, "Come on, Buddy, I'll show you a good place to eat."

"Just a minute," returned George, "I have something I must say to you."

Spooner put his grip down and said, "Go ahead."

George started slowly. "I am sorry that I let ya down, Mr. Spooner, I guess I am not big enough for the job. I couldn't keep that engine hot for you today and I am afraid I won't be able to do any better tomorrow. I wouldn't blame you if you turned me in. So in order to avoid any embarrassment for you I am handing in my resignation tonight."

After a few puffs on his pipe, Spooner took it out of his mouth and pointed the stem at George.

"Now, you listen to me and I'll put the record straight. It is I who should apologize to you," said Spooner. "That engine we had is in terrible shape. Here, let me show you her work report." Spooner went over to the clerk's desk and opened the report he had just finished. It read:

Flues leaking bad, boiler full of scale and valves out of adjustment. This makes the 7th time I have reported this engine. We lost 25 minutes tonight and I nearly worked my fireman to death. Have a good engine for me in the morning or get another crew because I refuse to take that cripple out again."

Turning to George, Spooner said, "There is not a fireman alive that could keep that leaky tea kettle hot. I'll be proud to have you with me in the morning, and I promise that you will never hafta go through another run like that again."

George felt a big burden roll off his back as they started for town. After supper, they each took a room at the Blackstone Hotel. Spooner requested the clerk to call them at 5:45 a.m.

There was only one tub on the floor and George was relieved to find it available. After a good hot bath, he went to his room, slipped under the blankets and tried to relax. He was desperate for sleep, but each time he closed his eyes, a vision of the pointer on the steam gage would fill his mind. The arrow always seemed to be aiming directly at him. What a situation, thought George, I fought that battle until I am thoroughly exhausted, and now it won't let me rest. Again he made an effort to relax. Gradually the scene shifted to the inner workings of that steam gage. There was that pesky little imp exerting all his might to hold the pointer away from the 200-pound marker. As he continued to study the rascal, an amazing transformation took place. The imp changed into a white angel with wings and

66

now the little creature was straining to push the indicator so that it would register 200 pounds. By the time the arrow was pointing straight up, George noticed a tiny corncob pipe sticking out of the corner of its mouth.

"A pipe-smoking angel, now there's one for the books," thought George. A vision or a dream? George couldn't tell which, but the last scene of that transformation helped to put his tired body to rest.

7.

Buddy Makes Good

Next morning, George met Spooner in the lobby and they walked over to a little place across from the depot. It was a quick order restaurant which had been converted from the shell of an old wooden diner. The windows along the side were covered with frost and flickering lights from numerous oil lamps created a crystalline pattern on each pane.

Molly, a heavy-set woman of about 50, operated the business by herself. Having catered to the railroad boys for many years, she knew most of them by their first names.

As Spooner and George sat down at the counter, Molly was doing the final mixing of her pancake batter. Apparently they were the first customers.

"Morning Frank, who's your young friend?" asked Molly as she wiped her hands on her apron.

"This is Buddy Williams, my new fireman," returned Spooner.

"Pleased to meet you," said Molly as she extended her hand over the counter, then placed two mugs of steaming coffee before them.

"Now! What'll you have boys?" she added.

"Make mine bacon and eggs," replied Spooner.

"That sounds good to me, I'll take the same," replied George.

A moment of silence followed during which both men appreciated several swallows of hot coffee. George started to speak.

"Mr. Spooner, do you —"

"Just call me Frank," interrupted Spooner, "Mister is reserved for gentlemen, of which I ain't. By the way, while we're on the subject, would you object if I continued calling you Buddy? Somehow that name seems to fit you."

"Not at all," replied George. "Since my performance yesterday, I was afraid you might be tempted to call me a few names less complimentary."

Spooner snickered and replied, "You did all right under the circumstances."

Spooner's determination to nickname him Buddy was soon taken up by the others, and George was obliged to answer to it for the remainder of his career.

"Here, boys," said Molly as she slid the plates of food before them, "This ought to last 'till you get to Milwaukee."

George sopped up every trace of egg with his toast. Spooner finished off his plate and started packing his corncob as George gulped down the last of his coffee.

Glancing at his watch, Spooner commented, "Guess we better get going."

While George was reaching in his pocket for some money, Molly studied his face. "Aren't you a bit young to be firing on a locomotive?"

Any reference to his age made George embarrassed and he hesitated to answer. Spooner tossed a half-dollar on the counter, "You don't need to worry about him, Molly, the only thing wrong is he was born too young."

Molly's face broke into a broad smile as she said, "Have a good trip, boys." George buttoned his coat up tight, slipped on his gloves, and picked up his grip. As they left, Spooner yelled, "See you tomorrow Molly."

A narrow path through the deep snow took them to the roundhouse. Spooner led the way. The grip in his hand seemed to be a part of him as he swung it in rhythm with each step.

The morning sun was about to make its appearance and a clear sky was getting brighter. Unlike the day before, there wasn't so much as a breeze stirring.

Walking by the huge freight yard, George observed that the switching crews had not yet started their day's work. A blanket of snow covered everything and the whole scene appeared serenely tranquil.

There is something about a freight yard that stirs the imagination. Hundreds of cars with varying silhouettes stand outlined against the horizon. Some are freshly painted with their different railroad emblems proudly displayed. A few new ones can be spotted here and there. Shabby looking, banged up gondolas loaded with scrap iron, are destined for Japan. Pulpwood piled high on flatcars is heading for the paper mills. Cattle cars crowded with restless steers are soon to be unloaded at the Chicago Stock Yards. Multiply this picture a thousand times and it becomes readily apparent why the railroads have been called the backbone of our nation.

After barging through the door at the yard office, Spooner went over to the clerk and said, "What engine have I got?"

"The ol' man assigned you the 1158," answered the clerk. They just finished a complete overhaul on her."

"Sounds like he received my wire," said Spooner, half under his breath. "Did Hoffman send any message for me?"

The clerk tossed him a yellow paper and said, "Read it yourself."

Spooner glanced at it and handed it over to George. It read, "GIVE SPOONER THE BEST ENGINE YOU HAVE AND TELL HIM, I'LL LOOK FOR HIM TO BE ON TIME TONIGHT."

"Come on Buddy, let's check out the replacement," said Spooner as he started out the door.

The thrill of walking through a busy roundhouse is a never-to-be-forgotten experience. A multitude of noises greets one as he approaches the building. Passing through the doorway these sounds become almost deafening. The ear-splitting staccato of a heavy rivet gun suddenly penetrates the very soul. The crush of a heavy hammer against a bulky ingot shakes the floor. A nearby grinding wheel scatters a shower of sparks, while the carbon steel screams out its objections to the abrasion. Each of these sounds seem to predominate a particular work area. Walking in front of a huge locomotive, the blast of exhausting steam gives one the impression that the iron monster is about to explode.

Spooner stepped in front of the 1158. She was gleaming with a new coat of black paint. After inspecting the valve gear, he remarked, "This is one of the newer models with the Walschaert valve gear. They are stronger on the start and stay in adjustment much better."

As George climbed into the cab, he noticed that even the interior was repainted. There was a good fire poppin' on the grates and the steam was about 125 pounds.

When Spooner raised the oil can out of the rack he shook it in a circular motion. Convinced it was almost empty, he removed the long spout and filled it with oil from the reserve tank. Before leaving the cab, he started the air pumps. George reached for the scoop. Bright blue paint covered the handle. Pulling it free he looked it over. It was a No. 12 that had never been used. Brakemen liked to kid firemen with, "All it takes to do your job is a strong back and a weak mind." That wasn't entirely true. The fact is without the lowly scoop shovel, not a wheel would turn and the whole railroad would shut down.

After one day on the main line, George was acutely aware of the value of a good shovel. The scoop slid under the coal pile with ease. Carefully he guided that first shovelful through the firebox opening. Before he turned for the second load, every muscle in his body sounded off. No doubt about it, a strong back was the first prerequisite for a fireman.

By the time Spooner returned, George was ready. "Our Departure is 6:55, let's head for the depot." George shut off the injector and glanced at the steam gauge. "I'm ready whenever you are Frank." "Give the bell cord a yank," returned Spooner, "that will alert the boys to open the doors." After the huge doors back of the tank were opened, the engine moved out onto the turntable and lined for the depot.

On the way, Spooner remarked, "I looked this baby over carefully and if they did as well on setting the valves as they did on the rest of her, she ought to be a dandy."

George nodded and smiled, but after his experience of the preceding day, the only thing that could put his fears to rest would be the way she steamed. As they eased down the track, Spooner opened the cylinder cocks to clear out the accumulated moisture; steam jetted sideways out of both ends of the cylinder. A blast escaped with each stroke of the pistons.

As the train was passing the depot, the brakeman hopped on the bottom step to the cab and said, "Morning, Frank."

"Good morning" returned Spooner, as he nudged the throttle further open. The engine moved ahead gingerly. "Have you a report on the snow conditions?" he added.

"According to the dispatcher," answered the brakeman, "we won't be bucking any snow today."

The brakie was a young man with a sharp looking new uniform. George compared his own baggy overalls with the brakeman's white shirt, black tie and cap, and he couldn't help but wonder how he might look in that outfit. When Spooner stopped, the brakeman ran ahead to throw the switch. While they backed down to the penstock, George started the bell. At one point in each swing, the sun would strike the shining brass and burst into a thousand rays. The engine looked brand new and George felt a strong sense of pride.

When they came to a stop, he lined the spout with the opening on the tank. While taking on the water, George moved over to the side near the depot to catch a view of the platform. Baggage and expressmen were busy loading. The train consisted of five cars: one baggage, two express, and two day coaches. George shut off the water, closed the lid on the tank and shoved the spout over in the clear. After coupling into the train, Spooner set the brake valve to charge the train-line.*

This particular operation seemed to trigger Spooner's need for a smoke. Filling his pipe was almost a ritual with him. First he would pack it down with his thumb and then test the draw. Often he would repeat this process two or three times before he was satisfied. After striking a match along his pant leg, he proceeded to light his pipe.

---

*This refers to filling the air brake reservoirs on each car.

As the conductor headed toward the engine, Spooner got off the seat box, squatted down in the gangway and received the orders from his outstretched hand. George busied himself making last minute checks: boiler half full, steam 198 pounds, extra coal in the back corners. Convinced that his preparation was complete, he sat on the seat box and wound his watch.

Spooner got the highball and answered with a couple toots. After releasing the air, (brake) he dropped the sand.* Turning toward George he asked, "Are you ready Buddy?" George gave him the highsign (twist of the wrist) and nodded. After taking the slack, Spooner heaved the bar over in the forward motion and opened the throttle.

Before the hind end of the train cleared the depot, George could tell that this engine was definitely stronger. He instinctively started heaving coal. However, after tossing in 10 or 12 scoops, he realized the steam was still right up there on the peg.** The injector was on and Spooner had the throttle in the last notch. Reluctantly he shoved the scoop back in the coal pile, sat down and waited for the pointer to start turning on him. As the train gained speed, Spooner started hooking back on the Johnson bar. The needle stayed right where it belonged. This was too good to be true. Here they were making around 40 mph. The steam never varied 5 pounds from the start, and part of the time he was sitting down. Keeping this engine hot became play and for the first time George really began to enjoy the new job.

Heading down a sag, the speed increased until the exhaust took on the familiar roar. Looking ahead one could see the little farming community of Wrightstown.

George moved over to Spooner's side and shouted, "This one is a good steamer." Spooner looked down at him, patted the handhold of the throttle and said, "She's a dandy all right. If the dispatcher leaves me alone, I'll have this train in Milwaukee on time."

---

*This expression refers to opening the valve which admits sand to the rail in front of the drivers.
**200-lb. mark

73

During the remainder of the trip, George found time to take in the beauty of the snow-covered countryside.

True to his word, Spooner drifted the last two miles and pulled into the Milwaukee depot on time. When he stopped the engine at the roundhouse, George removed his grip and commented, "By golly Frank, you were right, I didn't shovel half as much coal today."

Spooner knocked the ashes out of his pipe and said, "I watched the struggle you went through yesterday and I knew then you were going to make good. Today the steam never went back over five pounds on the whole trip. That's as good as anyone could ask for."

George continued firing for Spooner for over a month. What appeared to be a failure, turned out to be a valuable lesson.

Many firemen resented Spooner's method of running an engine. His obsession for being on time was interpreted as lack of consideration for his fireman.

When word got around that young Williams was firing for Spooner, the remark was passed, "Spooner will nail that young fellow's hide to the coal gate." Another comment was, "That poor kid ain't got a chance." Their prediction came dangerously close to fulfillment, but these men misinterpreted Spooner's motives. He was a dedicated engineer on a job which demanded more out of the engine than it could produce. Spooner was never lavish with his praise, but it wasn't long before the boys on the Lakeshore Division were all anxious to meet the new fireman who took all Spooner could dish out and came back for more.

When George was promoted to an engineer, he put to practice Spooner's philosophy. Right from the start, George gained a reputation for being one who demanded THE BEST from his engine and his fireman. Whenever it became necessary to work an engine beyond the endurance of his fireman, George would tell him to run the engine while he did the firing. This action was considered beneath the dignity of some engineers, but with George, it was just an opportunity to prove that he too had a heart.

Firing off the extra board is an excellent training ground for new firemen. They not only learn the road, but they also get acquainted with every type of locomotive. In this way, a fireman would often work for a different engineer every other trip.

Each engineer would contribute to the training program. George made a study of the operating methods of every man he fired for. These observations were filed in his memory to be used when he earned the privilege of taking over on the right side of an engine.

8.

Firing on the "Hotshot"

In the years immediately preceding the First World War, the nation was experiencing growing pains. The resulting industrial expansion followed the path of the rails.

The Lake Shore Division of the Chicago and North Western Railway threaded its way from Milwaukee to Green Bay, with several branch lines reaching east and west from Fond du Lac.

The Soo Line and St. Paul served much of the same territory and the competition was keen. In this area there was a series of huge mills which produced almost half the paper needed for the whole country.

The war scare nearly doubled the demand for newsprint, and the North Western responded with a new high-speed service, which promised shorter delivery time than its competitors.

The dispatcher assigned numbers 295 and 296 to the new run and it soon became known as the *Hotshot*. Right from the beginning, it was evident that everything had to click or else it would be impossible to make the time. The fast schedule put pressure on everyone from the superintendent to the section foreman.

Mr. Vilas, the general manager, had gained a fat contract on the strength of his promise to spot the paper at the Chicago warehouses early the following morning. All too often the phone on his desk would jingle and some angry voice would shout, "Where is my paper!"

Vilas' reaction would be to shake up everyone connected with the job.

The heaviest burden was carried by the engine crews. Running the locomotive on the *Hotshot* was no position for the fainthearted. In the first three months of operation, several experienced engineers found the schedule too demanding. Either they eliminated themselves or the dispatchers advised them to bid off, and to make room for a man who could handle the job.

Len Prunner had the reputation of being one of the best engineers on the division. He had enough seniority to hold just about any run he chose, but he thrived on challenges and the *Hotshot* appealed to him. Firing on this job was as rough on the fireman as the engineer. It was too much for the older firemen and many of the younger ones were afraid of it.

The train was hauled by the big Mikados with a 2-8-2 wheel arrangement — powerful freight engines with huge fireboxes.

Shortly after Prunner took the job, he started having trouble keeping firemen.

One evening, while making out his time card in the yard office, George approached Prunner. "How are things on the *Hotshot?*"

"I'm having trouble making the time without steam," replied Prunner. "These firemen they give me couldn't fire a flat iron out of a second story window."

"You know Len," returned George, "those Mikados are man killers and with that short time it's too much for a good man."

George had fired for Prunner while working off the extra board. As an engineer, Len was his ideal.

Prunner picked up his grip and started out the door. Stopping in front of George he said, "Frank Spooner told me that you could handle the job if anyone could. Why don't you bid it in and we'll show the boys how it's done?"

This was one job George hadn't fired and the prospect of understudying Len Prunner was very enticing. "I'll think it over," replied George as Prunner left the office.

77

Within a month George became the regular fireman on 295 and 296. The *Hotshot* proved to be the best classroom an ambitious fireman could have.

Night after night and in all kinds of weather, 296 entered the yard at Chicago on time. So consistent was their performance that the dispatchers would kid about setting their watches to match the arrival of 296.

On the return trip from Chicago, 295 hauled mostly empties. The lighter load made the fireman's duties considerably easier. They departed around 3:30 in the afternoon and except for two stops, they highballed all the way to Fond du Lac. Being away from the home terminal, they had to rely on sandwiches prepared in a restaurant near the roundhouse.

No. 296 would take siding at Clymon Junction for a Southbound passenger. During the wait, the engine crew would eat their lunch. The boys in the caboose could prepare their meals while en route and with the added facilities at their disposal, they fared pretty well.

One evening George almost choked while trying to masticate a gristle sandwich. In disgust he threw the remains into the coal pile. Looking over to Prunner, he said, "Len, I'm getting sick of these soggy meatless sandwiches."

"I know exactly what you mean," replied Prunner. "Sometime ago I complained to the restaurant owner, but it did no good."

Prunner had a couple of oatmeal cookies left from the lunch his wife had packed the day before. "Here Buddy try a couple of these."

"Len," said George, "I believe there's a way to beat this rap."

"Let's have it," returned Prunner.

"The other night," continued George, "I saw a follower plate° lying near the water tower at Janesville."

"I don't get you," said Prunner.

°A piece of iron about 14 inches square and 1 inch thick with a hole in the middle.

"Well, if I put that plate into the firebox about 10 minutes before we get into the hole,* by the time we get into the clear, it should be hot enough to fry a couple of steaks."

"Sounds great, but how do you figure on pulling that plate out of the fire?" inquired Prunner.

"I can snag it with a journal hook,"** answered George.

"By golly, it's worth a try. Here's a dollar for my share of the steaks," said Prunner.

The next time they took water at Clymon Junction, George put the follower plate on the tank and a friend in the car repair shop gave him a journal hook.

When he ordered his dinner, George told the waitress to fill the thermos but forget the sandwiches. Just before leaving the restaurant he casually slipped two forks into the pocket of his overall jacket, and mumbled under his breath: "This is part payment for those meatless sandwiches!" Then he went shopping: two thick T-bones, a bag of onions, salt and pepper, and a couple of pie tins.

When Prunner climbed up on the engine that afternoon, he looked over at George and said, "I don't have a thing but a thermos of coffee with me."

"Come over here and take a gander at this," said George.

"Now you're talking my language," returned Prunner enthusiastically.

Leaving Butler that night, Prunner seemed to be running a little faster than usual. While approaching Clymon Junction, George prepared a level spot directly inside the firebox door, and dropped the follower plate on it.

Prunner eased the train into the siding and George got out the steaks and trimmed off the fat with his pocket knife. After peeling the onions, he sliced them on the wrapping paper the steaks came in. When the engine stopped, he pulled the plate out of the fire and slid it on the apron of the tank. After throwing the fat on the plate, he moved it over the surface as it melted down. Prunner watched the whole proceeding with an amused expression. George seasoned the steaks and flopped

---

*Take siding.

**A service tool used to open journal box covers.

them on the smoking plate. The aroma of the sizzling beef filled the cab.

"How do you like yours done?" asked George.

"Just enough so I can grind it up with these store-bought choppers of mine," commented Prunner.

After turning the steaks over, George added the onions. The plate had cooled just enough to finish frying the steaks and still not scorch the onions. George handed Prunner a pie tin and a fork.

"Get your plate over here and I'll load it up," said George.

George started to scrape the onions to the side in order to push them on top of the steak. Prunner pulled his plate away unexpectedly with the result that some of the onions slid off on the deck. George did a slow burn as he gazed at the mess.

"Sorry about that," said Prunner apologetically. "I love onions but they hate me."

Sharing those onions was an act of extreme unselfishness. After swallowing hard, George covered his plate with the remaining onions.

Prunner picked the T-bone up in his hand and started gnawing on it. After his first bite he looked over at George and said, "This steak is as tender as a school marm's leg."

George decided to forgive him for wasting those precious onions. From that time on, Prunner always managed to take siding at Clymon Junction five to ten minues ahead of schedule. George kept a supply of groceries in the tank locker. The menu varied from steaks to chops to hamburgers and sometimes bacon and eggs.

Prunner bragged about George's "culinary" artistry until the head brakeman tried to work a deal whereby he could eat with them. But Prunner told him to go back and eat that sour stew with the rest of his crew.

Prunner and Williams formed a team known as the "dispatcher's fair-haired boys." George continued firing for him until he was promoted. The experience added greatly to the bag of tricks which eventually enabled him to "show the boys how it's done" while on the right side of the cab.

9.

The Engineer's Examination

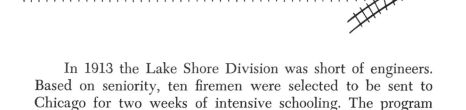

In 1913 the Lake Shore Division was short of engineers. Based on seniority, ten firemen were selected to be sent to Chicago for two weeks of intensive schooling. The program was designed to prepare the candidates for the Engineer's examination. As it turned out, George was the tenth one to be eligible.

While reviewing the credentials of each prospect, Steve Simmons, one of the officials, discovered a discrepancy in George's age. Promptly, he took the matter up with Herb Carkins.

"Do you realize that Williams is only 21 years old?" asked Simmons.

"Yes, I am aware of that," answered Carkins.

"You don't mean to tell me that you're going to turn that kid loose at the throttle!" exclaimed Simmons.

"There's no rule in the book that says I can't."

"All right," replied Simmons sarcastically, "but if he gets into trouble you'll have to answer for it."

"Matter of fact, Steve, I made a little investigation and found one instance where the North Western promoted a fireman who was only sixteen years old."

"I've never heard of such a thing," replied Simmons.

"You got any idea who he was?" asked Carkins.

"How should I know," snapped Simmons.

"For your information," added Carkins, "Williams' father, Charley, started firing for his dad at thirteen and was promoted when he was sixteen. I have been watching this young man, and I am convinced that he can handle the job."

After that last comment, Simmons stormed out of the office mumbling to himself.

With the hope of promotion in view, the idea of spending two weeks in the big city quickened George's pulse. He caught the noon passenger train and went home to prepare for the trip.

The whole family shared the excitement. Jessie pitched in and ironed his shirts. His mother replaced a few buttons and darned some socks. Ralph hung around and watched George pack his grip.

That evening after supper, Father Williams motioned for George to follow him out to the wood shed.

"You know, son," began his Father, "your Mother and I were hurt when Charley left home and failed to return. We had planned for him to follow me. Now we are pinning our hopes on you."

Brother Charley fired for a short time, but he suddenly left his job without explanation and skipped the country. Up to the time of this writing nothing is known of his whereabouts.

"Herb Carkins told me," continued his Father, "that there was some doubt as to whether or not you were old enough to be an engineer, but he persuaded them to give you a chance. Now you can be sure that they will be watching your every move."

His father paused for a moment in order to cram a chaw of tobacco in his mouth. "Carkins is one of my best friends, but I expect no favors from him. If you pull any foolish stunts, he will certainly wash you out. Have you got me?"

"I follow you, Dad," said George.

"One more thing, and I want you to remember this. Some engineers have never been able to properly interpret a time card.* They cannot be trusted on the main line. Others fail to operate within safe limits; these fellows do not last long. So don't take any chances. Whenever you plan to make a stop, always keep 200 feet up your sleeve."

*Schedule for train movements

"What do you mean?"

"In other words," said his father, "be able to stop at all times 200 feet before you have to. You can always drag the train to the exact spot."

"I'll be careful," promised George.

"If you study as you should for the examination, you won't have time to mess around. Chicago is full of hoodlums who are anxious to relieve you of your valuables, so stay out of the honky tonks."

George never gambled and he didn't care for drink. But he had packed a .32 revolver in his grip to deal with anyone who might try to rob him.

Before retiring that night, George slipped a copy of the book of rules in his coat pocket. Next morning he was on 206 heading for Chicago, with his head buried in the book.

George found a rooming house three blocks from the Chicago depot. It was a rundown three story brick building. His room was on the top floor with the bath at the far end of the hall. But the bedding was clean and a gas light was conveniently located over the table.

Early next morning he reported for class at the appointed room in the Chicago depot. There were forty students from four different divisions. The instructor passed out the study material. There were to be three separate tests. One on the book of rules, another on the Westinghouse Air Brake, and the third one was on the mechanics of steam power.

The North Western had been given several awards in the field of safety and the company was determined to provide each prospective engineer with the best training.

The next two weeks were filled with lectures, charts and diagrams. After the final tests were graded, George received a commendation for having the second highest score in the class.

The school was over by noon on Saturday, and George planned on heading back to Fond du Lac the next day. This left him with a free night. Some of the fellows in the class had visited an amusement center located near the stockyards. The attractions sounded interesting. Before leaving his room George stuck the revolver in his belt under his shirt. He was carrying $75 and had no intention of parting with it involuntarily.

A streetcar carried him right to the gate of the gayway. There were the usual thrill rides and sideshows. George went through the hall of terrors and took in several of the rides. It was getting late and most of the crowd had departed.

On a platform near the exit gate, George paused to watch a fire-eater perform. The Barker was encouraging the few remaining bystanders to "step right up and witness the once-in-a-lifetime experience" With a cane he pointed to large grotesque paintings of the different freaks. George paid his dime and entered the tent. After viewing each attraction, George started to leave. The barker stood in the exit and pointed his cane toward a dimly lit passageway. George looked around and noticed that he was the only spectator in the area.

"Step right here and view the unborn. Just one thin dime and you can witness a phenomenon withheld from the eyes of men until now."

Instinctively George suspected foul play. Cautiously he moved forward to go around, but the barker blocked the doorway by holding his cane waist high.

"I don't care to see the unborn," protested George. "I'll thank you to step aside."

About that time, George heard footsteps; looking around, he saw two fellows approaching him from behind. This confirmed his fears; he was being trapped. Slipping his right hand under his shirt, he pulled out the revolver and jammed the muzzle into the barker's mid-section.

"If your friends get any closer, you're a dead man." George spoke loud enough to be sure that the other two heard him.

"Now back out and be quick about it."

The barker dropped his cane and held his hands high. He appeared to stiffen in that position. The revolver gave a distinct click as George cocked the hammer. This sound seemed to re-activate his locomotion and he backed up hurriedly. George kept pressing the muzzle to his belly until they were way out into the street. With the gun still leveled at the barker, George started moving away. When he was satisfied with the distance between them, he made a dash through the gate.

Before selecting his seat on the streetcar, he walked from the front to the back and made sure none of the passengers resembled the barker. The next morning when George packed his revolver in the grip, he expressed gratitude to his Maker for deliverance. The thought of killing a man, even in self defense, gave him a sickening feeling.

"Just maybe," reasoned George, "I could have avoided this experience had I heeded Dad's warning." George's attitude toward his father was one of utmost respect and the happening that night only served to confirm the wisdom of his stern but loving father.

When George left Chicago on 209, he was in high spirits. During the stop at Milwaukee, he couldn't resist the temptation of walking up to the head end to see who was on. The relief engine was backing down to couple up. She was one of the new Pacifics, with the electric headlight. Frank Spooner was still the engineer.

George climbed up to say hello and tell Frank about passing the engineer's examination. Spooner grabbed George's hand and offered hearty congratulations. Turning to his fireman, Spooner said, "You think you got it tough? You ought to see what I put him through!"

George and Spooner had a good laugh. For some reason, the fireman didn't seem to appreciate the humor.

As George was climbing down off the engine, Spooner added, "Find yourself a comfortable position back there and I'll rock you to sleep on the curves."

As 209 left Milwaukee, George had a seat next to a window. While viewing the countryside, he listened for every whistle. The events of the day that he decided to quit railroading came back so clearly that he actually envisioned a repeat performance of the little imp who turned out to be an angel. At one point, he became so engrossed in thought that he started laughing aloud. The passenger facing him developed a puzzled expression which brought George back to reality with no little embarrassment.

As they were coming into Fond du Lac, George reached into his suitcoat pocket and pulled out a letter addressed to Herb Carkins, the traveling engineer. In it was a statement to the effect that George H. Williams had successfully completed the prescribed course, for the privilege of beginning his career as a locomotive engineer.

When the train stopped at Fond du Lac, George was the first one off and he headed for the baggage room. Picking up the Company phone, he called Carkins. "Hello, Herb, this is Buddy Williams; I have a statement signed by the Vice-President of the North Western which declares me eligible for promotion."

"That's great," answered Carkins. "Hold on a minute while I check the extra board." Shortly Carkins returned to the phone. "I marked you up on the board and you're five* times out."

"Thanks, Herb."

"By the way, where can we reach you?" asked Carkins.

"I'll be checked in at the McGivern Hotel."

"Don't stray too far from a phone," replied Carkins. "Looks like you will make your first run as an engineer tomorrow."

George caught the streetcar for North Fond du Lac and checked in at the McGivern. That evening before retiring he wrote a letter to his mother and dad, telling them the good news.   *   *   *

It has been said, "You can tell a hoghead by his walk and his talk." Every movement that he makes is deliberate. In any conversation, he speaks with authority. This facet of his personality stems from the fact that, on a locomotive, he is the boss. If the book of rules fails to get this point across, the attitude of the hogger soon erases all doubt.

The heavy responsibilities of an engineer demanded complete cooperation from his fireman. In the days of the hand-

*Refers to the fact that he will be the fifth in line to be called.

fired engine, the man with the scoop was at the mercy of the hoghead, consequently, treating his engineer with respect seemed to come naturally.

Due to an ill fitting curtain, the morning sun aimed its beam right into George's face. Awakening with a start, the first thing that entered his mind was, "Today I am the boss!"

Bouncing out of bed, he grabbed the towel, shaving equipment and headed down the hall to the bathroom. After getting dressed, he descended the stairs and stepped off into the lobby.

Sill Clausey, the owner and proprietor, was sweeping the floor. "Morning Sill, am I too early for breakfast?" he asked. Clausey straightened up and answered, "The wife has a good fire going in the cookstove, the coffee should be just about ready."

The McGivern Hotel catered exclusively to men. It was a three-story wooden structure with about sixty rooms. The whole operation was a family affair.

George went into the kitchen and gave Mrs. Clausey his order.

When he finished breakfast he moved into the bar and called the roundhouse. "This is Buddy Williams," said George. "How do I stand on the board?"

"Hold on and I'll check," he replied. "You're two times out," said the clerk. "Kinda looks as if you will get out of here this afternoon sometime."

"I'll be going downtown to get a haircut," answered George. "You can reach me at the McGivern after dinner."

As George replaced the receiver on the hook, Clausey leaned over the bar and stuck out his hand. "I understand congratulations are in order."

"Thanks much," returned George. "Maybe now we'll find out if I've learned anything," he added with a laugh.

After the haircut, George purchased a pair of gauntlet gloves and took his watch to the company jeweler. Enginemen are required to have their time pieces checked for accuracy once a month. The jeweler adjusted the second hand, screwed the crystal back on and signed George's inspection card.

Back in his room, George was polishing his shoes when someone knocked on the door. "Speak right up," he shouted.

"You're called for 3:00 to go west on a Sheboygan Extra," said the callboy.

"Thanks much," returned George, as he added a few more strokes with the brush.

With a quick glance at his watch, George started down the stairs. On the way, he picked up a paper in the dining room and continued on to the kitchen. Mrs. Clausey was peeling potatoes at the sink.

"I'll be leaving in about thirty minutes. Will you have time to fix me a lunch?" asked George.

"Beef sandwich all right with you?" she asked.

"Fine," replied George. "I'll be waiting in the dining room." With that, he sat down at a table and started reading the paper.

In a short time the cook came through the swinging doors and handed him his lunch. Thanking her, he went on in to the barroom.

"I'll have my usual," announced George, as he planted one foot on the brass rail. Clausey opened the icebox and brought out a quart of buttermilk. George removed the cap and proceeded to drink directly from the bottle.

The fellows in the bar would often kid him about his drinking habits. But George was convinced that buttermilk would cure everything from falling arches to failing eyesight. And besides, he loved it.

Returning the empty bottle to the bar, he picked up his grip with one hand, the lunch with the other and started for the roundhouse.

"Have a good trip," said Clausey, as George departed.

It was early in the summer and the wild grass was knee deep along the side of the footpath. The sun was shining and a few billowy clouds were hanging stationary in the sky.

George entered the yard office and signed the register. Looking up on the board he noted that Engine No. 479 was assigned to him. This type was called a "ten-wheeler" meaning she had four wheels as leading trucks and six drivers (three per side). Designed as combination passenger or light freight engines, North Western used them on their branch lines.

"You will find your engine on the north end of the round-house, near the coal shed," said the clerk.

"Who have I got for a fireman?" asked George.

"Freddy Kreiger," was the reply.

Approaching the engine from the side, George could see the white flag markers hanging limp on their staffs.* The fireman was busy when George climbed aboard. Shoving the scoop in under the coal pile, Kreiger pulled off his gloves and extended his hand. "Congratulations, Buddy." As they shook hands, he added, "You must be the youngest engineer on the division."

George thanked him and stowed his grip in the seat box.

Grabbing the oil can, George climbed down to lubricate the running gear. While finishing with the long spout, the brakeman came alongside. "Our train is ready on track 9," said the brakie. "Looks like about ten loads with seven or eight empties."

George continued up into the cab and took his position at the throttle. After wiping some oil off his gloves, he adjusted the lubricator and tested his injector. The fireman leaned out and checked both ways. "I'm ready whenever you are, Buddy."

"Start the bell," said George, as he made sure his side was all clear. The dingdong reverberated through the cab. George opened the cylinder cocks, pulled the throttle out five or six notches and the engine moved out.

The Sheboygan trains were made up heading north. It was necessary to back the train to Fond du Lac in order to pick up the branch line. At that point the train would be heading in the right direction.

After coupling onto the train, George released the air and moved the brake valve to the service position. The air pumps went into action and George watched the pressure build on the air gage. The conductor climbed up and handed over

*White flags must be displayed on the front end of all extra trains.

the orders. When George finished reading them aloud, the brakeman commented, "We have clearance on the main until 4:15. George checked his watch and added. "That ought to give us plenty of time to back down and get lined up for Sheboygan."

"What are we waiting for?" asked the conductor as he climbed off and started back. Before getting on board the caboose, he gave the highball and George started backing the train to Fond du Lac.

When backing a train beyond the yard limits, the conductor is in direct control of the movement. An air valve on the rear of the caboose operates the brakes. An air whistle is used to warn motorists at the crossings.

As the train continued past the Scott Street crossing, workers on the section crew stepped back and leaned on their shovels. When the engine moved on by, the youthful appearance of George's face caused a few second glances.

The head brakeman dropped off at the switch and George stopped as the pilot cleared. The brakie threw the switch and signaled to come ahead. George moved the reverse lever in the forward corner and opened the throttle.

Being slightly downhill for a half mile, the train started easy. By the time they emerged out into the countryside, they were traveling about 40 miles per hour. The Johnson bar was in the Company Notch* and the speed continued to pick up.

Spring thaws always produced soft spots in the subgrade and low rail joints became a constant concern. Track is laid so that the end of each rail falls approximately at the midpoint of the opposite rail. Section crews on the branch lines were spread so thin that keeping the joints tamped up to the proper level was a perpetual struggle.

At around 55 mph the engine began to develop a pronounced roll. George eased off on the throttle and she settled down.

Ten miles out of Fond du Lac, the playground of a country school lies adjacent to the North Western right of way. At this time of the day, children could be seen playing on the

*Positioned for the most economical cruising.

teeter-totters and swings. When George started whistling for a nearby crossing, the children gathered along the fence to wave. Like most engineers, George appreciated these friendly expressions. Waving back and occasionally adding a couple of toots was one of the nicer privileges of the job.

The fireman reached into the tank locker and pulled out the water jug. After taking a good swig, he sloshed out a cleansing splash and handed it up to George. About the time he reached out to take the jug, George spotted something moving in the middle of the track. It was nearly a half mile ahead and at that distance, it appeared like the sun's rays were shining off the back of a brown haired dog.

"Hold it, Fred," shouted George, as he grabbed the whistle cord and sounded several warnings. Situations like this were not uncommon. Usually, the animal got the message in plenty of time. When it became apparent that the object was not responding, he added a series of short blasts, but without results.

At this point, George felt an overwhelming sense of danger. Realizing that in order to stop in time, he would have to take immediate action, he slammed the brake valve into emergency, rammed the throttle closed, and flipped on the sander. While continuing to pump the whistle cord, one thought kept running through his mind: "I'll look pretty silly if some farmer's mongrel hops off the track as we draw closer."

The brakes were just starting to take hold when the truth became apparent. George's heart began to pound. There, seated in the middle of the track, facing away from the engine, was a little girl. What looked like the fur of a dog was the long flowing hair which covered her back down to her waist. Leaning forward, with her head down, the child sat there motionless. The action appeared to be a deliberate act of suicide.

Her position was such that even if the engine slowed down to 5 mph the pilot (cow catcher) would likely crush the child. Beads of cold sweat broke out on George's face, as he sat there helpless. After a quick analysis of the distance, speed, and braking action, George concluded that the engine would stop in time, providing the wheels didn't start sliding.

Finally, after what seemed to be an eternity, the engine came to a grinding halt. The pilot was less than ten feet from the child. Sitting there, with his left hand still clutching the whistle cord, George took a moment to relax and breathe a long sigh of relief. Uppermost in his mind was the question, "Why would this child want to take her own life?"

The sound of crying reached George as he leaped off the engine and hurried toward her. Stepping over into the middle of the track in front of the child, he spoke to her in a soft voice, "Little girl, why are you crying?"

Refusing to look up, she slowly shook her head and continued weeping. George removed his gloves, tucked them into his jacket pocket and squatted down. About that time he detected strange peeping sounds. Suddenly a tiny baby chick seemed to appear from nowhere. With a quick movement of her little hand, she returned the stray back under the shelter of her dress.

A huge lump came up into George's throat as he realized the reason for her action.

"We'll take care of your little friends," said George, as he picked her up in his arms.

A half dozen little chicks started to scatter. The fireman saw the problem and quickly lifted each one over the rail.

Screams from an hysterical mother broke the silence. Out of a farmhouse a short distance away, a woman came running.

"My baby! My baby!" she cried.

"She's all right, she's not hurt," shouted George as he hastened to meet her.

At a broken place in the fence, George returned the child to the trembling arms of her mother. After holding the child against her breast, she looked up at George with tear filled eyes and sobbed out, "I'm so thankful — so thankful."

George brushed his hand over the little girl's long brown hair and answered, "I'm thankful, too." As he spoke, he fought back a few tears of his own.

Turning on his heel, he started back. The conductor was all out of breath as he met George at the engine.

"What in the world happened?" he demanded.

George gave him a brief explanation and started to climb up into the cab. By laying a heavy hand on his shoulder, the conductor held him back.

"I came up here to bawl you out for being so rough with the brakes. Now I'm going back grateful for your action." Then with a broad smile and a friendly pat on George's back, the conductor took off for the hind end.

George sat down on the seat box and released the air brake.

"Hey, Buddy!"

George leaned out and looked back. About two car lengths away, the conductor stood with hands cupped around his mouth.

"I forgot to tell ya, you knocked the brakeman down with that fast stop."

"Leave him right where he fell," returned George, "I'll pick him up when I start."

The fireman and head brakeman started to laugh and George joined them. With five long and two short blasts of the whistle, he called in the rear flagman.

Looking over to his fireman, he waited for an "all clear." Kreiger checked on his side and returned with a "high sign." George made a couple of short blasts of the whistle and started the train. Looking back to be sure that the conductor got on board safely, George noticed that the mother was still standing in the field with her little one by her side. Both of them were waving. George returned the wave and responded with a toot-toot.

As the engine was gathering speed, George began to reflect on what had happened. One thing was certain. Had he delayed the braking action ten seconds longer, the child would very likely have been killed. The emergency application of the air brake can be very costly and sometimes dangerous. Engineers try to make sure that the need is truly an emergency. However, the time element is so critical that he must act almost instinctively.

As George relived the details that led up to the emergency, one thing became crystal clear: decisions of this order cannot wait for positive identification. And from this observation he determined to act as though all unidentified objects were human. In the years that followed, this decision spelled the difference between life and death for several individuals.

When George tied up at Sheboygan, the fireman came over to his side and said, "Buddy, you must have eyes like a hawk."

"How's that?" asked George.

"I'm supposed to have perfect vision," replied Kreiger, "but I would have sworn it was a dog in the middle of the track."

After a moment of silence, George adjusted his cap to the back of his head, looked down at his fireman and answered, "It wasn't my eyes that told me."

"I don't follow you," said the fireman.

"I really can't explain it, except to say that it must have been the hand of the Lord."

As George rested his head on the pillow that night, he saw his newly-acquired position as a locomotive engineer in a different light. From here on out, he would be making critical decisions and lives would be at stake. Somehow, the fact that he was the boss seemed to fade into insignificance.

10.

Lost, One Main Drive Wheel

Due to a critical shortage of motive power at the beginning of the first World War, locomotives were loaded beyond their designed capacities. At the same time there was a scarcity of skilled mechanics. Under these conditions, the steam engines suffered from lack of upkeep, and more than once this neglect resulted in a costly accident.

During this period, the Chicago & North Western was using Pacific type locomotives to haul their mainline passenger trains. These engines had a 4-6-2 wheel arrangement. Six-foot drive wheels (three per side) were connected to pistons with a 25-inch diameter. A pressure of 200 pounds per square inch delivered 4 power pulses for every turn of the wheel. Each revolution covered a distance of 20 feet. The manufacturer claimed that with seven steel coaches, on level track, this engine could accelerate to 60 mph within a distance of one mile.

No. 151 was a passenger run between Chicago and Green Bay. A Chicago Division crew handled the train as far as Milwaukee and men of the Lake Shore Division completed the run to its destination. Because the same engine was used for the entire trip, changing of the crews took place on the north end of the Milwaukee depot.

It was in July, 1918 and George had been running off the extra board. Working long hours in the hot humid Wisconsin summer takes its toll on sleep. This particular morning the telephone interrupted a much needed rest. "Hello, that

you, Buddy?" asked the callboy. "You're called to deadhead*
to Milwaukee in order to replace Bill Strang on 151," reported
the callboy.

"Okay, I'll be ready," replied George. Returning the telephone to the stand, he glanced at his watch on the dresser.
It was 5:30 a.m.

After a quick shave, followed by a hasty breakfast, George
packed his work clothes and caught the streetcar for the depot.

No. 206, the morning express, arrived on time and George
took a seat in the smoker. With legs crossed on the seat opposite and his hat pulled down over his face he managed to fall
asleep before the train left the station. The next hour and
twenty minutes would be devoted to getting even with that
infernal telephone. When the conductor came by to collect
the tickets he recognized George and decided not to disturb
his rest by asking him to produce a pass.

As the train pulled into Milwaukee, the brakeman lifted
George's hat and gave him a friendly smile. "Up and at 'em,
we're stopping for Milwaukee."

On his way to the depot George purchased a paper at
the newsstand and found a vacant bench in the waiting room.
It would be nearly an hour and a half before 151 was due
in, so George decided to read the latest war news.

When the huge clock on the wall reminded him that it
was time to get ready, George started for the dressing room
in the basement. After tossing his grip on the table, he noticed
fireman Raymond polishing his work shoes on the end of the
bench. Wallace Raymond was an excellent fireman, who had
worked with George on several occasions.

"What brings you up in this neck of the woods?" inquired
Raymond.

"I've been sent here to replace Bill Strang on 151," answered George, as he opened the grip and pulled out his work
clothes.

Raymond put the final touches on a good shine, straightened up and said, "Looks like we're going out together."

*Refers to traveling en route to the place of work. In this case it was from
Fond du Lac to Milwaukee.

"I think we're big enough for the job," returned George as he finished snapping on the straps to his overalls. Raymond checked his watch and said, "About time we get started."

Both men headed for the platform. As they arrived at the penstock, George opened his time card and reviewed the schedule. "Isn't it quite a struggle to make the time with all these stops?" he inquired.

"We used to get seven or eight cars out of here, but now they load us down with ten or twelve and it's darn near impossible," replied Raymond.

George pulled out his watch and started to wind it. "I see they are due in here in five minutes; are they usually on time?" he asked.

"You can figure they'll be at least five minutes late."

The main passenger gate in front of the depot was made up of vertical iron rods about ten inches apart. All passengers were kept back of this guard until the train arrived. A crowd was gathered in this area and the station agent was sounding off with the names of the cities along the route. Several wagons filled with express packages were lined up alongside the track. A switch engine was wheezing along with a mail car intended to go north on 151.

The familiar sound of a chimed whistle penetrated the air. Someone shouted, "Here she comes!" Several children crammed their heads through the bars to gain an early glimpse of the approaching train. Emerging around a bend about a half-mile south, the engine came into full view. The white exhaust from the stack was drifting lazily along the tops of the coaches and formed a sharp contrast against the background of dirty grey buildings.

While the engine was entering the shelter, George detected the distinct metallic clank of a loose main rod. When the train came to a stop, Raymond grabbed the long handle and swung the water spout over the tank.

Before climbing into the cab, George waited for the engine crew to get off. As he tossed his grip up on the apron of the tank, the departing engineer laid his hand on George's shoulder and said, "Better take up the wedges* on the right main rod and watch out for that reverse lever. This morning, leaving Chicago, the darn thing jumped the cog out of the quadrant and the bar banged hard into the forward corner. Good thing we weren't going fast," he added, "or we might have done some damage."

"Thanks much for the warning," replied George. "I'll keep an eye on it."

While Raymond finished taking water, the switcher backed down with the mail car, coupled it onto the train, and took off, with the bell sending out a noisy warning.

When the brakeman realligned the switch, George backed the engine up to the train. After setting the independent brake (engine brake) he started down with the oil can in one hand and a monkey wrench in his back pocket.

The long spout on the oil can was designed to enable the engineer to reach every moving part. George was a firm believer in making good use of it. While standing on his tiptoes he stuck the spout through the spokes of the huge main drive wheel and squirted oil on the lateral motion plates. Then he took a couple of turns with the monkey wrench on the screws which tightened the wedges on the main rod. As he returned the wrench to his pocket, he said to himself, "That ought to quiet down that noisy rod."

Moving around the front end to the left side of the engine, he aimed a couple of quick squirts at the piston rod and guides. Looking back, he could see Sampson, the conductor, coming toward him. The engine was located about 100 feet beyond the shed and the noonday sun was trying to nullify a faint breeze. George was filling the oil cup on the trailing trucks when Sampson came alongside. Breathing hard and with sweat dropping off his double chin, he handed the orders over to George. "Looks like we have our work cut out for us," he commented as he removed his cap and wiped the inner band off with his handkerchief.

---

*Four large screws on the main rod provided a wedge-type adjustment for taking up slack in the brass bearings.

Sampson was a big rotund full-blooded Indian. At first glance, children would almost become frightened at the expression of his broad wrinkled countenance. Actually, he had a strong resemblance to Sitting Bull, bowlegs and all. Being sensitive about his Indian blood, some of the railroaders soon learned that it was dangerous to joke about the subject. In spite of that, however, he was really a jovial fellow with a tender heart, and he liked nothing better than to put one over on a fellow-worker.

George put the oil can under his arm, removed his gloves, and read the orders aloud. A southbound extra would take siding for 151 at Campbellsport, and there was a 20 mph slow order for a mile north of the depot at West Bend.

"How many cars do we have?" asked George.

"We've got 12 and we'll be picking up a diner at Fond du Lac," answered Sampson, as he removed his watch from his vest pocket. George placed his timepiece alongside of the conductor's for comparison.*

"By golly, I see you're carrying around one of those wonder watches," quipped Sampson.

"What do you mean by that?" asked George.

"Why, every time you look at it, you wonder what time it is," returned Sampson as he slapped George on the shoulder and gave out with a hearty laugh.

Not to be outdone, George came back with, "You better sell that turnip you're carrying around on the end of that chain before the price of copper goes down. I see you're 20 seconds slow already."

The fireman was leaning out of the cab window and overheard the conversation. "That's telling him, Buddy," said Raymond.

"Oh, I see," said Sampson. "It's two against one up here. Think I'll go back among friends."

While climbing up into the cab, George hesitated long enough to scan the loading area. The passengers were all aboard, but one express wagon was holding up their departure.

---

*Engineers were required to compare watches with conductors to avoid any miscalculation.

After returning the oil can and wrench to the tray on the boiler head, George grabbed a piece of waste and wiped some oil off his gloves. The air pumps had been operating with a steady beat. Each oscillation of the compressors produced a noticeable jar that could be felt in the cab. When this movement ceased, George glanced up at the air gage for the main reservoir. The indicator was showing 130 pounds pressure.

Raymond had his fire in good shape, but the unexpected delay threw his timing off. The pop valves began to sizzle and threatened to let go. Leaning over from his seat he manually depressed the handle, which opened the firebox doors. Getting a little uneasy, George commented, "What's the matter with those fellows? Are they going to camp back there?"

Raymond leaned out and looked back. Sampson finally gave the highball as the expressman was pulling away his empty wagon.

"Let's go!" shouted Raymond as he reached over and closed the firebox door.

George released the air (brake) and backed the engine until he had the slack in four or five coaches. After dropping the sand, he threw the reverse lever over in the forward corner and pulled the throttle about one quarter open. When the slack was all out and the train began to move, he carefully opened it a little at a time. Before they had travelled a hundred feet he had her wide open. At one point, George sensed the big drive wheels starting to slip and he quickly rammed the throttle closed and jerked it right back out. At about 15 miles an hour George took hold of the brake valve and made a light application.* Satisfied that the response was normal, he quickly moved the brake valve into position for full release. When the resulting blast of air subsided, he returned the handle to the running position.

Going north out of Milwaukee was uphill and the long train held the speed down to around 35 mph. Black smoke belched upward with such velocity that a mushrooming col-

*When for any reason an air hose is parted the engineer is required to test the brakes at his first opportunity.

umn was forming nearly a hundred feet above the stack. In spite of the fact that George was getting all he could out of the engine, the fireman kept the steam right up under the pops.*

As the train reached level track, the frequency of the exhaust quickened noticeably. George shut off his injector and hooked her back three or four notches. Up until now Raymond hardly had time to straighten up and he welcomed the opportunity to relax on the seat box. After removing his gloves, he mopped off his brow with a big red handkerchief. George looked over at him and pointed at the steam gage. The arrow was right on the 200 pound marker. With a smile and a twist of his wrist, George indicated his appreciation for the fine way Raymond was keeping the engine hot.†

When it came to making up time, George described his own philosophy in this way: "Get 'em goin' and keep 'em goin'." Whereas most engineers would readily agree, very few would operate so as to fully utilize that principle. Some would do a good job getting the train rolling, but as soon as the speed picked up they would start reducing power. Others were not afraid to run at high speed but they took too long getting there.

Except in cases where the track conditions were subnormal, the North Western Rwy. insisted that their engineers observe a speed limit described in the rule book as "Safe and Prudent." Now it can be seen that this definition provided a lot of latitude for personal interpretation. There were special conditions such as curves, crossings, city limits, etc. But once George had a train out in the country and he was on short time,‡ the safe and prudent speed limit would somehow turn out to be the exact same speed he could coax out of the engine.

Eight miles out of Milwaukee, the North Western branches off in three directions. The speed limit at this junction is 40

---

*This expression refers to maximum steam pressure without unseating the safety valves.

†Or keeping the steam where it belonged.

‡Late.

mph. After cresting the hill it took another two miles to gain up to around 70 mph. About a half mile from the junction, George set the brake and watched the pointer on the air gage drop back 12 pounds. After lapping the brake valve,* he eased off on the throttle.

The speed was down to 45 as the engine passed the control tower. Looking back, George waited till the last coach cleared the crossover before he released the brake and opened the throttle. Instantly, the bark from the exhaust became louder and the train began gaining speed.

Coming into West Bend, George made a fast stop. The train crew hustled the passengers aboard. Leaving West Bend, the roadbed is slightly downgrade and the train started easier. George worked a light throttle until the engine passed over the track with the slow order. Then he pulled her wide open. The fireman opened the injector and busied himself with the scoop.

Seven miles to Kewaskum. It was a short stop and soon they were leaving a trail of black smoke. Another six miles and the engine leaned into a curve at the entrance of Campbellsport. George spotted a "45" on the arm of the semaphore.** Raymond shouted "45!" George raised his hand from the throttle and twisted his wrist. The wide cuffed gauntlet glove amplified the movement and assured the fireman that the engineer concurred. About 200 feet before the engine reached the signal, the arm of the semaphore swung straight up.

"Clear!" shouted Raymond, as he reached over and shut off the injector. George eased off on the throttle and shouted back, "All clear!"

Apparently, the extra had just pulled into the passing track at Campbellsport and the raising of the signal was the result of the brakeman realigning the switch for the main line.

Coming into the depot, the white flags on the front end of the extra could be seen fluttering in the breeze. The train

---

*Positioning the brake control to hold the existing application constant.
**A signal with a mechanical arm, the position of which provides information of the proximity of any trains immediately ahead.

was filled with soldiers on their way to France. The extra started pulling out before George came to a complete stop.

When the last passenger disappeared into the coach, the conductor gave the highball and George opened the throttle. It was 28 miles to Fond du Lac. With continued good luck, he hoped to pick up 3 or 4 minutes on the way.

About a mile out of town the roadbed started downgrade and the engine began gaining speed. Mindful of the warning given to him at Milwaukee, George took hold of the reverse lever in both hands and began moving it back a couple of notches at a time. As he squeezed the handhold to raise the cogs out of the teeth on the quadrant, he felt a jerking motion which threatened to pull the bar out of his hand. George figured that due to lack of lubrication, the piston valve was dragging on the cylinder wall. As long as the cogs remained seated in the teeth on the quadrant, the trouble was not serious.

Going by the little station at Eden, the train was travelling over 75 mph. From here on it was downhill almost all the way to Fond du Lac. The engine sensed the downgrade and continued gaining speed.

Because of unsymmetrical loading on the drive wheels, true counterbalancing could be achieved only at a certain number of revolutions per minute. These Pacifics would run quite smoothly up to around 75. However, the vibration at 85 became so pronounced that unless one bent his knees and stood on his toes, the action would actually give him double vision.

The roadbed had just been re-ballasted and the track was as straight as a bowstring. George counted on this stretch to regain precious time. Looking back, he could see the train was trailing straight and riding level. Strange as it may seem, many steam locomotives would actually run faster downhill with a ¾ throttle than with a full throttle. This phenomenon is due to the fact that beyond a certain number of revolutions per minute, the wide open throttle admitted more steam than the valves could exhaust. The resulting back pressure actually reduced the effective power.

George eased off to about ⅔ throttle. Going downhill, the heavy train was no longer a burden and the engine began to build up to a terrific speed and George just let her roll.

There was a steel girder bridge at the bottom of the hill with a curve about 600 feet beyond. Normally, George would apply the brakes about ½ mile from the curve. This day, he was going faster than usual and when he was a mile from the bridge, he took hold of the brake valve to start the application. At that instant, vibration caused the cog to bounce out of the teeth on the quadrant. The reverse lever became free to move and the dragging valve slammed it into the forward corner. The engine reared up on one side and there was a loud bang. The boiler began to pitch up and down violently and the cab began lunging from side to side. George threw the brake valve into emergency and braced himself so that he could lean out to see what happened. The main rod had torn loose from the crosshead and was gyrating around crazily, knocking off steam pipes, air tanks, running boards, and anything else that got in its path. A piece of piping sailed through the front window of the cab and George ducked back of the boiler head. In the midst of this pandemonium, George saw one possibility for survival. The lead trucks were still on the track and as long as they guided the engine there was hope. Glancing over to the fireman, George saw Raymond precariously balance in the window of the cab with his knees riding on the arm rest. It was 45 feet to the bottom of the fill and large rocks lined the banks. If he jumped at that speed it would be certain death.

"Don't jump!" shouted George. But his voice was completely lost in the crescendo around him.

Steam from bursting pipes began to fill the cab. George fought his way over to the fireman's side and grabbed Raymond around the waist. He seemed welded in that position. With a mighty jerk, George pulled him loose and they both rolled down onto the deck. The engine was still going over 80, and the continued jarring rendered them both unconscious.

Sampson, the conductor, was riding in the rear of the smoker. While finishing up a little bookkeeping, he observed that they had just passed the station at Eden. Noting the definite increase in speed, he commented to the brakeman, "If Buddy keeps this up, we'll be in Green Bay on time for a change!"

Suddenly the brakes went into emergency. The coach lurched. An elderly man walking in the aisle was thrown to the floor, and the brakeman helped him to his seat. Sampson peered anxiously out the window for some indication of the trouble.

The train finally came to a grinding halt. Sampson and the brakeman jumped off the coach and started for the head-end. Looking under the cars on their way, they noticed that every other tie was dislodged or broken in two. The hissing sounds of escaping vapors heightened their concern. The engine appeared to be enveloped in a cloud of steam. Looking up into the gangway Sampson cupped his hands around his mouth and shouted, "Hey, Buddy!"

The air in the cab began to clear, but from their position on the ground there was no sign of the engine crew. Sampson shouted again, "Hey, is there anyone up there?"

"They must have jumped!" suggested the brakeman.

"If they did, I'm afraid they're goners." replied Sampson.

The brakeman started up into the cab. "Here they are, Sam," he shouted as his eye caught the still forms of the two men.

George had a bear hug around Raymond's waist and they were both lying on their sides on the apron of the tank.

"My God!" exclaimed the conductor, "I think they're both dead."

At this point, George had come around just enough to hear the last comment. Opening his eyes, he looked up and said, "Where's Wallie?"

In astonishment Sampson kneeled down by his side and said, "He's right there in your arms."

The brakeman gave a sigh of relief.

It was necessary for the conductor to help unlock George's hands in order to release the fireman. Raymond started regaining consciousness. "Where in the world am I?" he said as he looked up into their faces.

George bent over him and said, "You're with me, Buddy Williams, and we just went through a wreck, but I think you're okay."

The brakeman helped Raymond to his feet. George and his fireman were so badly bruised that their faces looked as if they had been in a vicious fight.

"How do you feel?" asked George. Raymond remained silent and looked at him with a blank stare.

"Are you all right, Wallie?" added George.

Raymond shook his head as if to clear away some cobwebs. Suddenly his face brightened and he placed his hand on George's shoulder. "Buddy, you saved my life!" he exclaimed.

George choked up with emotion and couldn't respond for a moment. Tears filled his eyes and when his voice returned he said, "Kinda looks like the Man above was watching over both of us, Wallie."

The brakeman exclaimed, "We couldn't see you guys from down on the ground and we figured you both had jumped."

"I was going to," said Raymond, as he rubbed a large swelling over his eye. "I was just trying to get up enough courage when Buddy grabbed me from behind."

George found two large lumps on the back of his head and his body felt like someone had worked him over with a baseball bat. Fortunately neither had any broken bones. After giving a brief explanation, George suggested that they go down and look the situation over.

Before George stepped off onto the ground, Sampson hollered, "We lost a main driver!"

George walked over and examined the damage. Escaping steam partly obscured their vision.

"Here it is," said George as he pointed under the boiler. Apparently when the reverse lever let go, the valve adjusted itself for a full charge. At that speed, when the piston rammed against the trapped steam something had to give. The main rod transmitted the load to the main drive wheel and the force broke it off at the journal. This action tore the main rod off at the crosshead. The wheel derailed at once and being caged in by the rods to the other two drive wheels, started bouncing up and down between the boiler and the ties. After pounding a deep hole in the bottom of the boiler, it wedged itself up into the cavity.

After a short conference, Sampson sent the brakeman hot footing it back to the station at Eden with instructions to send for the wrecker and a relief engine. Meanwhile, the chief dispatcher notified the general office in Chicago that 151 passed Eden five minutes late, but was overdue in Fond du Lac by nearly an hour.

When a train makes an unscheduled stop, the engineer normally whistles out a flagman.* However, when there is an emergency stop, the hind brakeman assumes the need to protect the rear end.

As the brakeman trotted past the flagman, he told him just enough to let him know that no one was hurt badly. When he entered the depot at Eden, the agent asked, "What happened to 151? I've got every official on the Northwestern pounding my ear off."

The brakeman delivered the message and gave him a brief explanation of the accident.

Back at the yard office in North Fond du Lac, Steve Simmons read the accident report. Turning to the callboy, he asked, "Who's the engineer on 151?"

"Bill Strang is the regular man, but he laid off and Buddy Williams relieved him," answered the callboy.

"Is zat so?" replied Simmons, with just a hint of satisfaction in his voice. After calling the chief dispatcher for the particulars, Simmons picked up the report, walked into Carkins' office and tossed it on the desk in front of him.

---

*This is the signal for the rear brakeman to grab a red flag (by day) or a red lantern (by night) and run far enough behind the rear end to prevent any possible rear end collision.

"Looks like your boy Williams has got himself in real trouble," commented Simmons.

Carkins read the report and said, "What's the story?"

"Well, according to the agent at Eden, Williams went by flying low, and the engine came unglued about two miles north of there," answered Simmons.

"Did anyone get killed?" inquired Carkins.

"Williams and his fireman were shaken up a bit, but the engine stayed on the track."

"What makes you so sure that Williams is at fault?" said Carkins.

Simmons leaned over Carkins' desk and banged his fist down hard. "Five will get you ten that the little s.o.b. was going over a hundred miles an hour!"

"Better be careful what you say, Steve," suggested Carkins. "Buddy might make you eat those words."

The phone rang; Carkins answered. The chief dispatcher gave him instructions to prepare a work train with the wrecker and proceed to the accident as soon as possible. Then he added, "Is Steve Simmons around there?"

"Hold on," replied Carkins, as he handed Simmons the phone.

"Hello, Simmons?" said the dispatcher. "I want you to ride the relief engine behind the work train and bring back the disabled engine."

The work train arrived on the scene in less than two hours. After seeing the extent of the damage, Carkins remarked, "It's just a miracle she didn't leave the rails and tip over."

George gave Carkins a rundown on the cause of the accident. Raymond added, "That's the truth, Herb, I saw the reverse lever when she let go."

The wrecking crew went right to work and within an hour the relief engine was coupled up and the work train was on its way back with the wrecked engine.

Carkins offered to take over for the remainder of the trip, but George assured him that he was all right and that there would be no need for him to ride beyond Fond du Lac.

As is the case in all accidents, the North Western made a thorough investigation. It just so happened that the Chicago man on 151 had reported the engine for the same trouble on two separate occasions. The testimony of George and his fireman verified the cause. At the close of the meeting, George was completely exonerated.

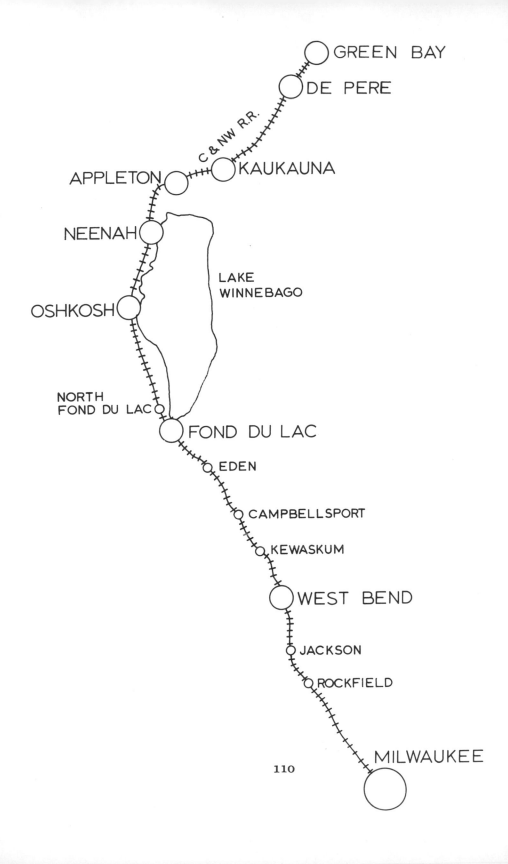

GREEN BAY

DE PERE

C & NW R.R.

KAUKAUNA

APPLETON

NEENAH

LAKE
WINNEBAGO

OSHKOSH

NORTH
FOND DU LAC

FOND DU LAC

EDEN

CAMPBELLSPORT

KEWASKUM

WEST BEND

JACKSON

ROCKFIELD

MILWAUKEE

110

11.

Racing the Soo Line "Mountaineer"

The Soo Line gave special attention to details that were often overlooked by the bigger companies. Their rolling stock was maintained in the finest condition. Considerable effort was made to keep their engines clean. Even the grounds around their roundhouses and shops were beautifully landscaped.

Although their primary source of income came from freight revenue, the management showed real pride in their passenger service. *The Mountaineer* was one train that was particularly outstanding. It was an all-Pullman limited which ran from the Canadian border to Chicago. The train crew dressed in smartly tailored uniforms, and a Soo Line emblem was neatly woven in gold on the lapels of their coats. The engineer and fireman wore white high crowned caps and freshly laundered overalls.

Originally this train was pulled by an Atlantic type locomotive and consisted of four Pullmans and one baggage car. However, its popularity grew until eight cars were required to accommodate the passengers. The schedule was fast and the added tonnage became too much for the engine, so the Soo purchased a bigger locomotive to handle the heavier train. The new power was a handsome high-wheeled Pacific. She had a graceful taper to her boiler and the steel tires on each of the drive wheels were painted white.

By comparison, this engine was approximately the same size as the North Western's Class "E" Pacifics. Although the

111

Soo engine had slightly smaller cylinders, she made up for it by having a higher working steam pressure.

From the outskirts of Neenah to North Fond du Lac, the Soo's main line runs parallel to the Northwestern. Except for a short distance through the city of Oshkosh, the two rights of way are less than one hundred feet apart. This thirty-mile stretch of track provided an occasional opportunity for a good contest.

It was in the fall of 1927 and George had been working off the engineer's extra board. This particular afternoon, he was called to run on a switcher at North Fond du Lac.

From the McGivern Hotel to the roundhouse, the footpath crosses over the Soo's main line. On his way to work, George was stepping over the rails when he heard the low pitched moan of a Soo whistle. Looking north to the top of the hill, he saw the *Mountaineer* rocking along toward him at high speed, and paused to watch the train go by. When it drew closer he saw that the engine was one of the new "hi-steppers." * As it roared on by with whistle wide open, George commented to himself, "That baby is a going machine!"

Starting back on the path, George met Harry Madison. Apparently he had just completed a run and was on his way home. Madison spent many years on mainline passengers and George considered him to be one of the fastest engineers for whom he ever fired. Not having seen him for a long time, George greeted him and stuck out his hand. "Did you notice how fast that Soo Line passenger was traveling?" he asked.

"Sure did," replied Madison, "and if those engines run uphill as well as they do downhill, I doubt if we have a locomotive that can stay with them."

"Remember the time I was firing for you on 216 and we met a Soo passenger leaving Oshkosh?"

"How could I forget?" Madison replied with a chuckle. "That's one time we gave the Soo Line engine crew a long look at the rear end of our train. By the way," he added, "what engine did we have?

---

*Fast engine with tall drivers

"Seems to me it was the 1616," answered George.

"You must be right," said Madison, "I had her on 102*
the other night and she's still the best engine we have."

George glanced at his watch. "Guess I better be gettin'
along, I'm due on duty in five minutes; see ya around!"

About two weeks later while George was having supper
at the McGivern Hotel, someone shouted, "Hey Buddy! the
callboy wants you on the telephone." George hurried to the
barroom and picked up the phone. "Start the conversation,"
said George as he swallowed his last mouthful.

"You are to deadhead to Green Bay and come out of there
on EXTRA tomorrow," said the callboy.

"What's the occasion?" inquired George.

"The Packers squad and coaching staff have chartered a
train for Chicago," answered the callboy.

"What time are we scheduled to leave Green Bay?" asked
George.

"12:30 p.m."

"I'll be there," replied George, as he hung up the receiver.

Having been an admirer of the Packers football team,
George was looking forward to pulling their train. Before retiring that night, he notified the callboy to wake him in time.

Early the next morning George headed for Molly's Diner,
just across from the south end of the Green Bay depot. The
morning sun was busy erasing the evidence of a heavy dew
and the air smelled especially fresh.

While approaching the crossing, the warning bell started
dinging and the long wooden gate swung down across the
road. George watched as a switch engine labored to shove a
string of passenger cars onto a sidetrack in front of the depot.
Suspecting that this would be his train, George noted the
number and type of cars. There were two chair cars, four day
coaches, and two baggage cars — eight cars in all.

As the gates swung back up and the switcher cleared the
crossing, George picked up his grip and walked over to the
diner. The tantalizing aroma of bacon frying reached him as

---

*102 was a southbound passenger.

he entered the door. Moving to the far end of the counter, George sat down and started reading his newspaper. Molly had been busy at the cash register and didn't see him enter. On her way back to the stove, she recognized him. Reaching over she tipped up the bill of his cap until it rested precariously on the back of his head.

"Good morning, Molly. I'll have bacon and eggs," said George.

After cutting off several thick slices of bacon, she slapped them on the grill. George sipped his coffee and started reading the sports section. The Green Bay paper devoted a full page to the Packer-Bear football game. "Looks like the home team is going to have their hands full tomorrow."

Molly eased a couple of eggs onto the grill. "They better win, everybody in Green Bay is betting on them," she replied. "What job are you going out on?"

George answered, "I'll be leaving here at 12:30 on the *Packers' Special.*"

"Be sure and give our boys a nice smooth ride," replied Molly.

"We'll take good care of them," promised George, departing.

On his way to the roundhouse, George stepped along briskly. He greeted the clerk and after studying the bulletin board for a moment, asked, "Who have I got for a fireman?"

"Pedeletzski," was the reply. Pedeletzski was one of the younger firemen but he had learned quickly and in a short time was regarded as one of the best on the division.

"That's encouraging," replied George. "I'll have a heavy train out of here and I'm going to need all the steam I can get."

A voice in the locker room shouted, "Sounds like you're going to bend me over that shovel all the way to Milwaukee."

Unknown to George, Pedeletzski was in the adjacent room changing clothes. Looking over to the clerk, George came back with, "My fireman is complaining about being overworked and he hasn't even stepped foot on the engine yet. Looks like we'll meet 151* at Oshkosh," he remarked as he folded the time card and stuck it in the rear pocket of his overalls. "Guess we better get going." George picked up his grip and started for the door. Pausing in front of the clerk's desk, he asked, "What engine do I have?"

"Here," said the clerk, as he slid a yellow paper across the desk.

George read the message aloud: "PULL ENGINE NO. 1616 OFF OF 210** AND GIVE IT TO ENGINEER WIL-LIAMS ON THE *PACKERS' SPECIAL.* SIGNED, WALTER HOFFMAN, MASTER MECHANIC."

"You should have seen Engineer Duncan blow and storm when he found out they were giving you his favorite engine," commented the clerk.

Tossing the paper back on the desk, George addressed Engineer Duncan as though he were in his presence. "My thanks to you, Mr. Duncan, be assured I'll take good care of your pet." As he finished the sentence he bowed slightly as though he was begging leave to depart. At the conclusion of the little act, all three men had a hearty laugh. George turned to his fireman and said, "Come on, let's get that engine on her way to the depot before Hoffman changes his mind."

Each steam locomotive had a personality of its own. Even though one engine was the exact blueprint of another, their characteristics would differ considerably. So many variables contributed to the overall performance that even the experts couldn't explain the difference. Once in a great while an en-gine would accumulate more than her share of plus factors. When this happened the locomotive would be exceptional, and 1616 was a good illustration. She could pull stronger, run faster, and steam better than any of her sisters.

On their way to the back of the roundhouse George com-mented, "Duncan treats his engine like a mother hen does her one chick. No doubt he'll worry about it until he gets her back."

---

*151 northbound passenger train.
**210 southbound passenger train.

The 1616 was being moved off the turntable onto the lead for the depot. The hostler* stopped the engine in front of the water tank and climbed off on the engineer's side. "She's all yours, Buddy," he said as he started back to the yard office.

Slipping on his gauntlet gloves, George climbed into the cab. Before stowing his grip he opened the firebox door and inspected the crown sheet. A thin layer of glowing coals was spread evenly over the grates and the steam gage indicated 125 pounds. Glancing back he could see the fireman pulling the waterspout down into the tank. Beginning with the gage cocks, George started his routine checks. Reaching forward alongside the boiler, he opened the primer on the injector, and when the sound indicated water was moving into the boiler, he shut off the primer and screwed the injector closed.

The fireman came over the coal gate and started building up his fire. George glanced at his watch and decided he had sufficient time to complete the oiling before leaving for the depot. Slipping the monkey wrench into his pocket, he grabbed a piece of waste in one hand, the oil can in the other, and climbed down to the ground.

The rods on the 1616 were lubricated with a very heavy grease which came in the form of round sticks about six inches long. These sticks fit into a cylindrical reservoir located near the crank pin in the rod. By inserting the grease in the hole and screwing a heavy steel plug down on top of it, the lubricant was forced into the bearing area. This operation was usually performed by the machinist before each run. However, a supply of these sticks would be carried on the engine for emergency use. Being one of the most vulnerable spots on the running gear, engineers would usually give a couple of extra turns to the plug on each rod, just to be sure.

George squirted oil on the journals and lateral motion plates and with the wrench he took a couple of turns on each grease plug. Moving toward the front end, he applied a generous stream to the piston rod and crosshead guides. Reaching overhead, he placed a quick squirt on each bearing in the

---

*Hostler — One whose job it is to move engines in and out of the round-house.

valve gear. Before heading around the front end, he squatted down and inspected the flanges on the pony trucks. This protrusion on the inside periphery of each wheel extends a scant two inches below the rail. Its function is to guide the locomotive on the track. As this 150-ton mass of iron thrusts its weight against the outer rail of each curve, a grinding, scuffing action removes part of the material. In time the wear can weaken the flange so that it may fail, and cause a disastrous wreck. When he finished oiling, George was satisfied that he had properly prepared the running gear. Climbing back up into the cab, he turned his attention to the feed water lubricator which was an ingenious device anchored near the top of the boiler head. Resembling a Scottish bagpipe, its function was to meter oil to the cylinders and air pumps.

A built-in reservoir held about a gallon of oil. Three valves on the bottom controlled the rate of flow. By means of sight gages, one could visually observe the oil drops fed into the pressure-filled lines. As the throttle was opened, the oil was atomized into the cylinders. The prescribed rate was one drop for each two or three seconds. In order to be certain that it was full, George closed the steam valve to the reservoir, removed the filler plug and stuck his finger into the opening. It was full.

This kind of oil was expensive and the company frowned on using more than recommended. But George belonged to the old school which taught, "If a little will help, a whole lot will cure." High speed increases the danger of piston seizure; therefore, George always doubled the oil flow and considered it a wise investment.

A quick glance at his watch told him it was time to get going. "Let's head for the depot," said George as he released the independent brake. When the fireman gave him the high-sign, George opened the throttle. The 1616 barked sharply as the wheels made a full revolution before the engine started. Three hundred feet down the track a ducted shield on both sides of the rail provided a safe place to open the blowoff cocks.

One of the lessons George picked up from Prunner was the importance of keeping the boiler free of scale. The North Western's engine's were notorious for being over-cylindered. This meant that during periods of maximum demand, the boiler was not big enough to maintain the steam pressure. Consequently, anything which improved the steaming capacity, increased the available power.

"Give her a good shot," said George as he stopped the engine in the proper location. A two-inch jet of boiling water blasted out on each side with a roar like the sound of a dozen pop valves letting go. George watched the level drop in the water glass and shut off just before it went out of sight. Spewing out 300 gallons of hot water may seem wasteful but the cleansing action more than compensates for the loss.

The fireman put on his injector and tossed in a half dozen scoops of coal.

"Start the bell," said George as he cracked the throttle.

A small valve on the engineer's side controlled an air driven piston which oscillated the bell. But it required a pull on the cord to get it started. The track was in need of repair and the joints were particularly bad. Even at low speed, the engine sensed the dips and started swaying. George eased off on the throttle and let her drift.

On the approach leading to the house track, George spotted his brakeman giving him the "come along."* When the engine pulled on by, he hopped onto the bottom step to the cab and signalled to continue on.

The switch engine, with a diner in tow, passed them at the crossing, going in the opposite direction.

"Where are they going with that?" asked George.

"They are hooking it on the hind end of our train," replied the brakeman.

While crossing over onto the main line, the brakeman dropped off, aligned the switch and gave a backup signal. As the engine moved on by, he hopped onto the back of the tank and rode until they coupled into the train.

---

*Hand signal meaning to come ahead.

Soon as the air hose was connected, George moved the brake valve to the service position.* The air pumps responded by giving out with their peculiar rhythmic beat.

Several wagons piled high with football gear were being unloaded into the baggage cars. Suddenly three large buses backed into the parking area, directly across from the engine, and spewed forth the men of the Packers' team. Each man was neatly attired in a suit. Except for their size, one would hardly guess that they were some of the nation's greatest athletes.

The men gathered on the platform directly below the cab and a newspaper photographer was busy flashing pictures. One of the big fellows looked up at George and hollered, "How's chances me riding up there with you?"

George leaned out the window, pointed to the steps leading up into the cab and said, "Be my guest."

The fellow made a move toward the cab when another big bruiser grabbed him by the arm and pulled him in the direction of the coaches. Looking back he shouted, "Some other time?" George smiled and nodded his head.

The conductor was making his way toward the engine and as he drew near George recognized Donnahue. Climbing off to meet him, George pulled off his glove and said, "How are you, Jack?"

The men had mutual respect for each other. The experience of that scrap in the Neenah yard actually drew them together as friends.

"The officials are particularly anxious to have this engine in Milwaukee in time to come back on 209," said Donnahue, as he handed George the orders.

"I don't see any reason why we can't accommodate them," replied George as he scanned the orders and handed them up to the fireman. The Southbound Extra was to take siding on the passing track at Oshkosh until 151 cleared.

"After we take water at Fond du Lac, we can highball straight through to Milwaukee," said Donnahue. With that he

*This position would start the air pumps charging the trainline.

pulled out his watch. George placed his Elgin alongside. It was 12:25 p.m.

"Soon as we get through loading, I'll give you the high-sign," returned Donnahue as he started back.

George climbed back up into the cab and took his position at the throttle. The working pressure of the 1616 was 180 pounds. The steam gage was indicating 178 and the pop valve began to sizzle. Pedeletzski reached over and opened the fire-box door to cool her down.

The last few pieces of baggage were loaded and the head brakeman gave the all clear. The rear brakeman repeated the signal and Donnahue gave the highball. George shouted, "Highball!" dropped the sand, moved the brake valve into full release and reached for the throttle. The 1616 leaned into the heavy load. After two or three exhausts, she threatened to stall. George pulled her wide open and the train started. Before moving a full car length the drivers threatened to slip. George shut off and jerked her back out about halfway. At around 15 mph he took hold of the Johnson bar and hooked her back four or five notches. The sharp crack of the exhaust was like music to George's ears. Obviously the valves on this engine were in good shape.

When the rear end of the train cleared the crossing, George made his running brake test. As soon as he sensed the retarding action, he released the air and returned the brake valve to the running position. As the speed increased, the crossings kept him busy with the whistle cord.

Nine cars was a heavy train for the 1616 but once it got them a'going, she could run with them. George let her roll through the little towns of De Pere, Wrightstown, Kaukauna, and Little Chute. The North Western cuts through the center of Appleton and several crossings present constant danger. George slowed to around 35 mph as they entered the city. Leaving town he widened on the throttle. Coming into Neenah, the signal for the St. Paul crossing was clear. George eased off and drifted through town at around 35 mph. As he finished whistling for the main crossing by the Valley Inn, George pulled open the throttle.

Emerging into the country, George noticed a farmer ploughing in a nearby field. The man waved and George gave him a couple of toots. Almost like the echo of his own whistle, two answering toots came back at him. Looking forward, George spotted the rear end of a Soo Line passenger. It was a quarter of a mile ahead, running parallel. From the lazy way the smoke was rolling out the stack, George concluded the Soo Line man was laying back to entice him into a race.

Through the years the engine crews of both railroads developed an unwritten set of rules. If there were the prospect of a friendly race, the engine in the lead would limit his power until the challenger could gain sufficient ground to equalize their chances. When this point was reached the engineer in front would signal and start with two blasts of the whistle. Being against the rules, racing was strictly unofficial. However, a spirit of competition would permeate the whole crew. If the contest became particularly close, it was not uncommon for the passengers to shout encouragement much the same as spectators at some sports event.

George caught Pedeletzski's attention and pointed ahead. When the fireman realized the situation, he clapped his hands together like an elated child. Then, glancing over to the steam gage, he noticed the pressure had dropped back eight or nine points. Immediately he reached forward on his side of the boiler and shut off the injector.

Each additional pound of steam pressure added over fifteen horsepower to the engine. Pedeletzski grabbed the scoop and started stoking the fire.

George had the throttle wide open and the 1616 was performing beautifully. But just to be sure the engine was producing every ounce of available power, he leaned out and analyzed the sound of the exhaust. In a few seconds he determined that she could stand another notch forward on the reverse lever. When he finished making the adjustment, George listened to the bark from the stack. A smile spread over his face as he detected an improvement.

Black smoke blasting skyward from the Soo Line engine was evidence that they intended to stay out in front.

121

The fireman managed to build the steam back up to 175 pounds and the needle was still climbing. Moving over to the gangway on the engineer's side, Pedeletzski noted the Soo engine was pulling eight cars. A round marquee on the back end of the last car spelled out, "The Mountaineer." "It's *The Mountaineer!*" he shouted. "Do you think we can take 'em?"

Leaning over, George spoke directly into his ear, "We're pulling more tonnage, but if you'll get that needle on the 180 pound marker and keep it there, I'll try to put the rear end of this train past them before we get to Oshkosh." He finished the sentence with his index finger pointing at the steam gage. The message was clear. Pedeletzski took hold of the scoop and continued firing with renewed vigor.

Both trains were accelerating, but the 1616 soon closed the gap and started moving up. As he passed each car, George saw that the passengers were crowding the windows.

Conductor Donnahue was relaxing in the forward coach when the rear end of *The Mountaineer* came into view. It was not long before the Packers were shifting over to the right side of the cars in order to watch the progress.

As the speed of both trains increased, the advantage enjoyed by the North Western began to diminish. Just as the 1616 was pulling even, the roadbed started on a series of rolling hills. Downgrade, the Soo's engine would start creeping ahead, but the 1616 would gain it back on the climb. The contest was so even at this point that both engines were whistling for the approaching crossings in perfect unison. The low pitch of the Soo Line whistle blended perfectly and the result was a pleasant harmony that echoed over the countryside.

Recalling the comment made by Madison, George thought to himself, "If this race were all downhill, *The Mountaineer* would likely be the victor."

Midway between Neenah and Oshkosh, a gradual climb topped out at a flag stop called State Hospital. When the weight of the heavy train began to be felt on the hill, George adjusted the reverse lever two notches forward. The resulting increase in valve travel produced a bigger charge of steam in the cylinders and held the speed at around 65. *The Moun-*

taineer began to fall behind. As the little station of State Hospital slipped by, George looked back. The rear end of their last coach was a car length in front of the Soo Line engine.

George motioned for the fireman to come over and see the results. "Take a gander at that," he shouted, pointing to the rear. Pedeletzski took hold of the grab irons and leaned out. Spouting a constant jet of black smoke, the Soo Line engine was still struggling to stay in the race. Leaning over toward George, Pedeletzski shouted, "Let's keep 'em back there!"

A short distance out of State Hospital, the roadbed started downward and it continued its gradual descent for another two or three miles. Soon it became apparent that *The Mountaineer* was coming back strong. George glanced at the steam gage. The pressure was back four or five pounds. "Get with it!" he shouted. The fireman shut off his injector, jumped down on the deck and started bailing coal.

Even though the 1616 was accelerating, *The Mountaineer* was slowly moving up. Two miles outside of the city limits, the two railroads diverge. Both trains were rolling close to eighty miles per hour when the cow catcher on *The Mountaineer* was directly across from the tank on the 1616. In a matter of seconds, George would have to start the braking action or run the risk of an emergency application.

As the two roadbeds returned to the level, the relative position of each engine remained constant. Realizing that when he shut off and set the brakes, *The Mountaineer* would shoot past him, George was reluctant to take the action. The Soo Line engineer was in a similar dilemma. Having made a remarkable recovery, the possibility of winning seemed to be within his grasp.

George took hold of the throttle, depressed the latch and hesitated. Suddenly the advice of his father came back to him. "Remember to always keep 200 feet up your sleeve." Immediately, George shoved the throttle to a drifting position, grabbed the brake valve and made a 15 pound reduction.

Fully expecting *The Mountaineer* to take the lead, George was relieved to see the smoke rolling off the stack of the Soo Line engine. Apparently the two engineers made the same decision simultaneously. Just before the brakes started taking hold, *The Mountaineer* hit the curve which took it on a diverging path through Oshkosh.

The switch leading to the passing track was a quarter of a mile north of the Oshkosh depot. Approaching the siding, the speed was down to 10 mph. George was preparing for a full stop when a brakeman at the switchstand started giving him the "come along." Frequently, when a switching crew is in the vicinity, one of the brakemen will align the switches for an oncoming train. This little courtesy allows them to ease into the siding without stopping. George thanked him with a toot-toot, released the air brake and cracked the throttle.

The engine drifted down to the other end of the siding and George climbed off and walked over in front of the main driver. Removing a glove, he held his open hand close to the center of the wheel. After making certain that the hub was not radiating excessive heat, he touched the area with the palm of his hand. George repeated this process to each drive wheel. When he returned to the cab, the fireman was taking a drink from the water jug. "Here, Buddy," he said, handing him the jug. "Pour some water on my hands." George had been so taken up with the race that he overlooked the tremendous effort put out by his fireman. The back of his jacket was soaked and beads of sweat were rolling down his face. George shook some water into his fireman's hands. "Stand back," he said, as he threw the water in his face and rubbed it around the back of his neck. Holding his hands out he said, "Let's try that again." This time he swished the water over his curly hair. As George watched, his memory carried him back to the days when he was firing for Len Prunner on the *Hotshot*.

"Would you believe it," commented George, "I checked the drivers and the journals were hardly warm."

"After a hard run like that, it wouldn't surprise me if one of them caught fire," replied Pedeletzski.

"If we had had another mile downhill those Soo Line boys would have gone by us like the pay car passed the tramp," said George.

No. 151 could be heard whistling as she entered Oshkosh. George looked at his watch, "Not too bad, we've been here less than ten minutes." After 151 passed, George gave two tugs on the whistle cord and looked over to his fireman. Pedeletzski gave him a twist of the wrist. George threw the reverse lever over in the backup position and cracked the throttle. After three or four exhausts he moved it back in the forward motion and pulled her about half open. Starting out on the main line through a spring switch, the *Packers' Special* began picking up speed. Pedeletzski had the bell ringing and George gave a low whistle for each of the many crossings. Homes line both sides of the track going south out of Oshkosh and out of consideration for the people George was light on the whistle.

The signal for the bridge at South Oshkosh was clear. "Clear!" shouted Pedeletzski.

"Clear!" returned George as he widened on the throttle. The 1616 was making close to 40 mph when she rumbled over the bridge.

As the train continued to gather speed, George pulled out his watch and studied it for a moment. Looking out the window he observed the rate at which the poles were passing (before speedometers, some engineers calculated speed this way). They were going about 60 mph, and barring any unforeseen delays, George concluded that he would arrive in Milwaukee in plenty of time. Out of consideration for his fireman, he decided to hook her up a couple of notches and ease off to half throttle. The exhaust quieted down noticeably and the train continued rolling at the same speed. Pedeletzski welcomed the relief as he got up on the seat box and took a breather.

Three miles south of Oshkosh the Soo's main line converges toward the North Western's track. George was leaning out the window watching the rail ahead when the sound of a Soo whistle penetrated the cab. Scarcely a half mile away, *The Mountaineer* could be seen closing in fast. Judging from the angle and speed they were moving, George could see that they would meet side by side.

Pedeletzski had just finished stoking his fire and was in the process of scraping up a few chunks of coal that had spilled out on the deck. Quickly he rammed the scoop under the coal pile and stepped over to the gangway. *The Mountaineer* was rounding the curve which guided it alongside.

"Do you think those boys want another crack at us?" shouted Pedeletzski. Two sharp blasts of the Soo whistle answered his question almost before he finished the sentence. "Give 'em your answer, Buddy, and we'll do it up right this time," added Pedeletzski.

Sensing his fireman's enthusiasm, George was only too glad to respond. With two quick jerks on the whistle cord, he answered their challenge and pulled the 1616 wide open. *The Mountaineer* was traveling four or five miles faster when they met and the Soo Line engine was obviously going all out. Consequently it continued right on by until the people in the last coach were waving goodby to George. At this point, however, the 1616 managed to hold its position.

While leaning out to evaluate the sound of the exhaust, the pop valves let go. Pedeletzski was on the deck bailing coal. George decided she could stand one more notch forward. The speed continued to increase and the rolling of the engine was accentuated. George took a wide stance and carefully moved the reverse lever forward one notch. The fireman cut on the injector and the pop valves settled down. George glanced at the steam gage. The pointer was right on 180 pounds. "If I lose this one, I certainly can't blame my fireman," thought George. The last adjustment was bringing results and the 1616 started fighting its way. George couldn't resist extending an occasional wave.

Both roadbeds start upgrade about eight miles south of Oshkosh. The two engines were exactly even at the start of the climb. George was counting on this section of track to put the *Packers' Special* past *The Mountaineer*. Both trains began to lose speed as the climb became steeper. George dropped her down a couple of notches. The 1616 proved to be the stronger and before they leveled off, the last car of the North Western's train was right across from the Soo Line engine. When the train passed over the top, the speed started picking up. George hooked her up two notches. Looking back he saw *The Mountaineer* tenaciously holding its position. Just before passing Tower DX the track started downgrade, with Fond du Lac less than five miles away. The Soo Line train started bearing down. George realized he would somehow have to squeeze more speed out of the 1616.

Pedeletzski moved over to the gangway and looked back. When he saw the situation he turned to George and said, "Buddy, do something!" George focused his eyes on the water gage — the glass was over ¾ full. High water level in a boiler tends to produce foaming action at the throttle valve. This action puts moisture in the steam and reduces the expansion, resulting in loss of power. "Shut off your injector," shouted George, "and open your blowoff cock!"

Both men held the blowoffs for about sixty seconds and the water level dropped to ¼ of a glass. George decided to try hooking her up one more notch. The combination added just enough power to check the advancing *Mountaineer*. The speed was close to 90 mph and the 1616 seemed to vibrate all over.

Feeling somewhat relieved, George leaned out and sighted down the track. About a half mile ahead, something caught his attention. In a matter of seconds he realized that a herd of cattle had wandered onto the right of way and was in the process of crossing the track. George held the whistle wide open. The cattle appeared totally oblivious to the warning. George tried striking the whistle with a series of short blasts. They continued on at the same pace. George started to reach for the brake valve but the folly of trying to stop or even slow down soon became apparent. At that speed, nothing short of an emergency application would even start taking effect before the impact. The Soo Line engineer saw the situation and added his whistle to the warning.

A huge bull in the center of the track lowered his head and started for the engine. "Take cover!" shouted George as he slammed his side window closed and ducked back of the boiler head. "BOOM!" a sound resembling someone striking a huge bass drum was heard above the roar of the exhaust.

Looking out the side window, George saw the mangled carcass sailing past the window. Returning to the seat box he found the forward window splattered with debris and the deflector smeared so badly that he couldn't see through it.

George pulled down on the bill of his cap and leaned out into the wind. The 1616 was approaching the yard limits at North Fond du Lac. While holding the whistle open he took a quick glance toward the rear. *The Mountaineer* had dropped back considerably and the smoke from the engine was barely clearing the stack. Apparently they had given up the race.

George eased off on the throttle and made a 15 lb. brake reduction. By the time they passed over the Scott Street crossing, the train was slowed to around 60 mph.

The fireman moved over to George's side, "How close did they get to us?" he asked.

"Their front end never got within six car lengths of our engine."

"Say, I'll bet that bull got quite a surprise when we hit him," said Pedeletzski with a laugh.

"I'm afraid we must have hit two or three of them," returned George as he added another five pounds in the brake cylinders.

The train passed in front of the depot going about 15 mph. George released all but ten pounds of air and let her coast the last one hundred feet. The 1616 came to a gentle stop with the tank lined up for the penstock.

A pungent odor filled the cab. "Smells like the Chicago stock yards," remarked George as he removed the oil can from the rack. Conductor Donnahue met him with the orders as he stepped to the ground.

"Buddy, you ol' son of a gun, tell me, just how fast will this engine run?" Donnahue followed George as he continued oiling.

"There was a short while when I feared she wasn't quite fast enough," said George as he poked the spout through the spokes.

"Things got kinda quiet back there in the coaches when *The Mountaineer* started gaining. What happened?"

"We had a little high water so we blew her off and I hooked her up another notch. That seemed to do the trick," replied George.

"You should have heard the cheer that went up as *The Mountaineer* began to fade." The comment really appealed to George and it brought a broad smile to his face. "By the way," added Donnahue, "what was all the whistling about?"

"We had a little collision just before we got to North Fond du Lac," answered George as he aimed a squirt at the crosshead.

"From the smell around this engine, I'd say you hit a loaded manure spreader," said the conductor. Donnahue followed George to the front end. The remains of a cow were jammed up onto the pilot and the whole area was well decorated.

Donnahue took one look and exclaimed, "We can't pull into the Milwaukee depot with a mess like that!" The position of the animal was such that it appeared as if a good tug would make it slide off.

The conductor grabbed one horn in each hand and pulled. The body started rolling off with unexpected ease. In his attempt to step back, the conductor's heel became hung up on a track spike. George made an attempt to catch Donnahue and pull him clear but the limp carcass continued to roll and part of it landed on the conductor. Donnahue raised up on his elbows to size up the situation. After one glance he lay back and just groaned.

George didn't dare smile — much less laugh. Donnahue gave out with a few choice words and started to extricate himself. "Hold it, Jack!" said George, "Let me help." With that he took hold of the horns and rolled the carcass off to one side. The conductor sprang to his feet, pulled out his handkerchief and brushed off a few spots. As it turned out, his uniform suffered very little from the experience.

"We've got rights over everything between here and Milwaukee. You can take out whenever you're ready," said Donnahue as he started back.

George crossed over to the other side of the engine, applied a couple of well-aimed squirts and climbed back into the cab. Pedeletzski had finished taking water and his fire was ready.

"What do you say?" asked George as he took his place at the throttle.

"I say, let her flicker," returned the fireman.

George released the air and reached up for the whistle cord. Two quick toots echoed back from the adjacent buildings and George opened the throttle.

George began working the engine hard in order to get a run for the hill. They hadn't traveled a half mile when both injectors were required to keep water in the boiler. As the fireman took hold of the scoop, George looked down at him. He was a mass of perspiration. As George hooked the reverse line back a couple of notches, he felt a twinge of conscience. The engineer creates the need for steam but it is the backbreaking labor of his fireman that produces it.

In the last thirty miles, Pedeletzski had more than earned his day's pay. As he took hold of the shovel, George stepped down on the deck, tapped him on the shoulder and said, "Get up there and run this engine for me." A puzzled expression came over Pedeletzski's face as he pushed the scoop back into the coal pile. "I've been needing some exercise and now's as good a time as any to get it," added George.

The fireman climbed up on the engineer's seat box, pulled out a big red handkerchief and mopped off his brow. George grabbed the scoop and started firing.

Young firemen always appreciated the opportunity to run a locomotive. In this way they gained valuable experience. George frequently extended this courtesy. This day, however, his object was to give him a well deserved rest.

The remainder of the trip was uneventful. After uncoupling at Milwaukee, the 1616 proceeded to the roundhouse.

The two men walked over to the yard office and made out their time slips. George asked the clerk for an accident report form. "Here ya go," replied the clerk, as he tossed a pad on the desk.

George wrote a short report on the accident, folded the paper and dropped it into a slot. Turning to Pedeletzski he said, "Come on, let's get washed up." Both men took a quick shower and put on their street clothes.

The office phone rang just as George picked up his grip and started to leave. "Hold it, Buddy," yelled the clerk. "Did you just come in on the 1616?" he asked.

"That's right," answered George.

"Here, the roundhouse foreman wants to talk to you," said the clerk as he pushed the phone toward him.

Grabbing it up, George said, "What can I do for you?"

"Let me see now," said a voice on the other end. "I'd like to have you send up twenty pounds of calves liver, sixty pounds of stew meat, and . . ."

"Wait a minute," interrupted George, "What do you think this is, the local meat market?"

"Well, after taking one look at the front of the engine, I'd say it looks more like the slaughter house," returned the foreman. "How in the h - - - do you expect me to get this engine cleaned up in time to go back on 209?" he demanded.

"I guess you got a problem there," replied George. "And by the way," he added, "better do a good job or Engineer Duncan might get a little rough with you."

A string of profanity spewed out of the receiver until George put it back on the hook. Pedeletzski was holding his sides and laughing to kill himself.

"Let's get out of here before somebody else tries to get on my back," said George as he grabbed his grip and barged out the door.

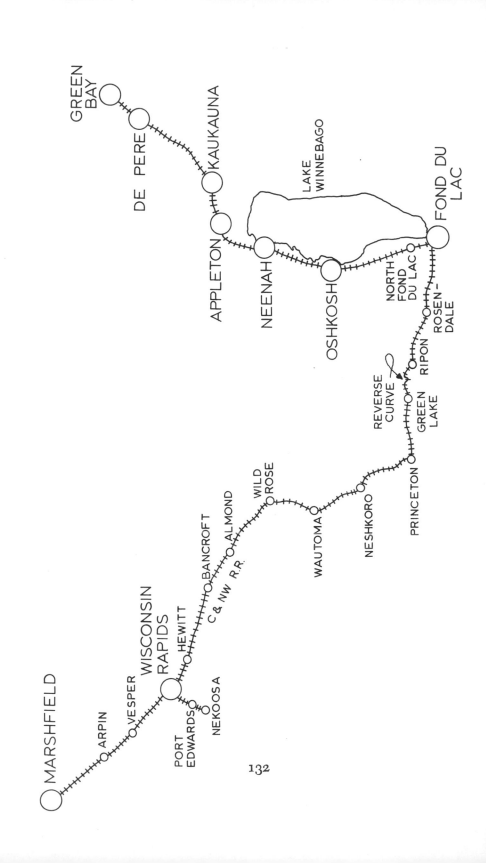

GREEN BAY

DE PERE

KAUKAUNA

LAKE WINNEBAGO

APPLETON

FOND DU LAC

NEENAH

OSHKOSH

NORTH FOND DU LAC

ROSEN-DALE

RIPON

REVERSE CURVE

GREEN LAKE

PRINCETON

WILD ROSE

ALMOND

BANCROFT

NESHKORO

C & NW R.R.

WAUTOMA

MARSHFIELD

WISCONSIN RAPIDS

HEWITT

VESPER

ARPIN

PORT EDWARDS

NEKOOSA

132

12.

Snowbound

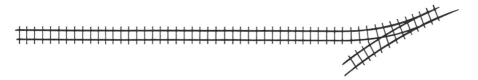

Geographically, the location of Wisconsin gives no indication as to the severity of its winters. The ground turns white in December and stays that way until March.

As late as 1930, automobiles and trucks were handicapped by the lack of all-weather roads. In times of emergency, such as floods or blizzards, the smaller towns were almost totally dependent upon the service provided by the railroads.

The North Western right of way cuts through hundreds of rolling hills. Snow, driven by the winds, skims over the ground and eventually deposits itself on the lee side of every trough.

On the main line, where the traffic was frequent, drifts would usually be destroyed before they could accumulate. However, the passage of one or two trains a day was insufficient to keep the tracks clear, so the branch lines were more susceptible to snow hazards. The development of the rotary plow did much to alleviate the problem, but those rigs were expensive and all too often the need was so widespread that there were not enough of them to meet the demand.

It was during the first week in January of 1928 and heavy snow had been falling steadily for 24 hours. Buses were stalled and highway traffic came to a standstill. Those who were traveling were doing so by rail. The depot at Fond du Lac was crowded with passengers awaiting the arrival of the northbound limited. The wind was beginning to pick up and lights on the platform accentuated the swirling action of the huge flakes. Occasionally a gust would reach sufficient force to produce a low pitched hum in the weatherstripping around the main entrance.

Leaning over his desk the station agent cleared some frost from the window. Across three sets of tracks, dimly outlined by the lamp at the intersection, No. 9, the westbound passenger, was almost obscured by the falling snow. This train operated on a branch line between Fond du Lac and Marshfield and travelers from the north and south would make connections at Fond du Lac for the west end. Due to the fact that Ripon was a college town along the route, a high percentage of the passengers were students.

Christmas vacations were over and an extra coach was added to handle the anticipated increase. There were six wooden cars in all — four day coaches, one baggage, and one combination mail and express. On the head end was one of the North Western's famous high-wheeled Atlantics.

Before the advent of steel coaches, this locomotive steamed proudly in front of the North Western's finest trains. Though she lacked pulling power, when it came to speed, she could still outrun most of the new engines.

The *Wisconsin Special,* No. 209, was due in about ten minutes. Outside, baggagemen and expressmen were jostling loaded wagons into position. In the waiting room an ambitious paperboy was threading his way among the passengers shouting: "Get your *Commonwealth Reporter* here!" Above the incessant chatter, the clickity-clack of the telegraph key was clearly audible. Inside the ticket office a telegrapher was busy copying down an incoming message. At the conclusion, he tucked the pencil back of his ear and tossed the paper over in front of the station master. "Kinda looks as if the officials are getting worried about the conditions on the west end," he remarked. The agent pulled out his glasses, adjusted the light over his desk and read aloud. "Deliver the following instructions to trainmaster Thompson arriving on 209: Recent reports indicate hazardous conditions north of Ripon — you'll ride No. 9 and render assistance as necessary. Signed Jack Rice, Division Superintendent."

"I'll see to it he gets it," said the agent as he slipped the paper into his pocket.

About that time, a noisy buzzer started sounding off. This was an indication that 209 had passed the cross-over at the tower south of town. As he reached over to shut it off, the agent could see the yellow beam of the headlight in the distance. Hurriedly he buttoned his coat and started pulling on his gloves. Walking through the waiting room, he announced, "Train Number 209, the *Wisconsin Special,* now arriving on track Number 2."

As he stepped out on the platform, the train was approaching the depot. Snow was packed under the smokebox and the pilot was completely covered.

The trainmaster was among the first to get off. Hurrying to meet him, the agent held out the paper and said, "Here's a message from the superintendent."

Moving back under one of the lights, the trainmaster read the instructions. "I guess there's no rest for the wicked," he commented. Thanking the agent, he picked up his grip and joined the passengers transferring to No. 9.

Trainmasters are usually picked from the conductor's roster. Having proven themselves to be superior in their field the company relies on these men to troubleshoot problems related to train movements.

MacDonald, the conductor, was helping passengers with that first big step and he didn't notice the trainmaster standing by his side. When the last person climbed aboard, Thompson tossed his grip up on the vestibule platform and greeted the conductor. "Howdy Mac, I'm afraid you'll have to put up with me this trip."

"Glad to have you with us," returned the conductor. "Could be I'll need all the help I can get," he added as the trainmaster disappeared into the coach.

A blast of air came from the engine on the 209 and the bell went into action. Soon the bark of the big Pacific was reverberating from the surrounding buildings and the northbound limited was headed for Oshkosh. As the blinking taillights faded in the distance, MacDonald slipped his hand inside his coat, took out his watch, and held his kerosene lantern alongside. "Ten minutes late, that's not too bad," he said half aloud. Looking fore and aft he could see his two brakemen were ready. "All aboard!" he shouted as he swung his lantern in a big arc. The head brakeman repeated the signal, and the engineer responded by releasing the air and starting the train.

The train gathered speed along the roadbed which was covered by a 6-inch blanket of white. The cowcatcher had been replaced by a low profile plow. Cascading snow formed a continuous arc as it was being scooped up and tossed to one side. These conditions would have caused real concern for an inexperienced engineer. But Al Hassman had spent many years running over this pike and he knew the route like the palm of his hand.

❋   ❋   ❋

The dispatcher's office being the nerve center of the division is a busy place even when the railroad is running smoothly. A steady stream of information flows in via telephone and telegraph. This data is evaluated and translated into orders. In a very real sense the dispatcher calls the signals much like the quarterback on a football team.

When one considers the problems involving as many as thirty trains on ten sets of rails, many of them moving toward each other at varying speeds, you begin to appreciate the job performed by a dispatcher. While determining routes and schedules so as to avoid collisions, he must also consider the most efficient way to deliver the goods.

The chief dispatcher sits in his little glassed-in cubicle and keeps a jaundiced eye on the whole proceedings.

In spite of the clatter of telegraph keys, the ringing of telephones, and a thick pungent cloud of tobacco smoke, these men go about their monotonous tasks with proficiency. However, when there is an emergency, the panic button gets mashed and the tempo immediately doubles. A derailment, for instance, may affect only one route and therefore the consequence is limited. But when a snow storm hits, the area involved is usually so widespread that paralysis threatens the whole division. It becomes the responsibility of the chief dispatcher and his assistants to untangle the mess and return the operation to normalcy.

It was about 7:30 p.m. and the men in the dispatcher's office were in the process of finishing their lunch. (As a mutual consideration agents try to minimize the sending of messages during mealtimes.) Leaning back in his swivel chair, with his feet crossed on his desk, the chief was enjoying the first puff of a fresh cigar. A telephone started ringing and continued for an extended period. Finally, with a note of irritation in his voice, the chief shouted, "Answer that phone before it drives us all crazy."

One of the clerks picked up the receiver and announced, "Dispatcher's office."

"What in the h --- is the matter over there, have you all taken the day off?"

The voice was so loud that the clerk developed a pained expression and held the receiver away from his ear. "Get the chief on the phone," he demanded. By this time, everyone in the office recognized the caller as Jack Rice, the superintendent.

Dropping his feet from the desk with a thud, the chief removed his cigar and grabbed his phone. "Chief speaking, what can I do for you?"

"Give me the latest report on the west end passenger," said Rice.

The chief snatched a yellow paper out of his "in" basket. "No. 9 departed Fond du Lac 10 minutes late," he replied. Glancing up at the clock on the wall, he added, "We should be getting a wire from the agent at Rosendale any minute now." A telegraph key began giving out with a rapid staccato. The telegrapher turned toward the chief and nodded his head. "Hold on, Mr. Rice," replied the chief, "We're getting something on it now." Soon the key became silent and the telegrapher handed over the message. While adjusting his bifocals the chief leaned over and started reading into the mouthpiece. "Here ya go. No. 9 arrived Rosendale 8:30, departed 8:40." After jotting down a few figures, he added, "I calculate they lost 15 minutes. Could be the snow is slowing them down."

"That's just what I am worried about," returned the superintendent. "Send off a wire to the agent at Ripon telling him to get a report on snow conditions from the engineer on No. 9." The chief was busy composing the message when the superintendent finished speaking. "Have you got that straight?" inquired Rice.

"I got ya okay," returned the chief as he anchored the pad to the desk with his elbow and peeled off the top sheet.

"Now if conditions get any worse, I want to send out a plow from North Fond du Lac and . . ."

The chief interrupted, "Sorry, Mr. Rice, every plow on the division is being used."

"How about Ole Granny* up there at Wisconsin Rapids?" asked Rice.

"That's a pretty ancient piece of equipment and I hate to take a chance with it," replied the chief. "Besides, we've got only two engines up there and both are being used on the switch runs."

Everyone in the dispatcher's office was taking in the conversation.

"How soon will there be one available?" asked the superintendent.

Glancing at a chart on his desk the chief studied it a moment and replied, "We sent out three plows this afternoon and I won't have one back until tomorrow night."

A moment passed in complete silence. The chief spoke up. "Mr. Rice, I'd like to make a suggestion. There is a reverse curve just north of Ripon and at the rate the snow is shifting, those cuts are bound to be filling up. Even if No. 9 manages to get through, we should have a crew out there cleaning up those bad spots." The chief paused and blew a perfect smoke ring.

"Get on with it," snapped the superintendent.

"If you can get the master mechanic to furnish us with a good engine, I'll start the switching crew putting a work

---

*"Ole Granny" was a wooden gondola loaded with fifty tons of rock. Anchored on one end was a flying wedge type iron plow. Having served the area for nearly twenty-five years it was about due for the scrap heap.

train together. The roadmaster can round up all available section hands and load the tool car with enough shovels for everyone. With good luck they can be on their way by 11:30. Does that make sense to ya?" concluded the chief.

"Under the circumstances, I can't think of anything else that does make sense," replied the superintendent. "While you're at it, better order out an extra crew to go along. No telling how long they may be fighting that snow. If any emergency comes up," continued the superintendent, "you can reach me at my home."

"Will do," returned the chief. With that he hung up the receiver. "Send this wire off at once," he said as he leaned over the desk and slid the message through a slot provided for this purpose.

✿   ✿   ✿

Leo Washneck, the clerk in the yard office at North Fond du Lac, lost his left arm and part of his right hand due to an accident while working as a fireman. The railroad gave him a cash settlement and a job for life. In spite of his handicap, he proved to be an efficient worker. Pounding away with his one good finger, Washneck was laboriously typing up new cards to be mounted on the bulletin board. An engine crew just returning from a run was changing clothes in the adjoining locker room. The topic of their conversation was the recent problems they encountered while bucking snow.

The weather door slammed and someone could be heard stomping snow off his feet. With the earpiece on his cap pulled down, a wool scarf wrapped around his neck and wearing a heavy Mackinaw, Bill Langland barged into the office. He removed his cap and slapped it against the bench. Looking over at Washneck he commented, "This kind of weather spells nothing but trouble. Have we had any problems, so far?"

"No, by golly, things have been strangely peaceful." With that he made a few determined pecks on the typewriter, pulled the card out of the roller, and proceeded to insert it on the board.

Stepping into his office, Langland removed his coat and hung it back of the door. Glancing at an engine report left on his desk, he shouted, "By the way Leo, can you tell me if they finished installing the plow* on the 479?"

"The machinist stopped in just before going home tonight to tell me he got the job done," replied Washneck.

"Good," returned the division foreman, "we may need her before the night is over."

Walking over to the window, Langland peered through a clear spot on one of the panes, and saw toward the back of the roundhouse that the snow was moving in a diagonal pattern. The faint sound of an engine's exhaust, mingled with the metallic clank of a loose main rod, could be heard in the distance. Somewhere out in that miserable weather a switching crew was faithfully doing its job.

Washneck was busy filing some reports when the company phone rang. Slipping the remaining papers under the stump of his left arm, he picked up the receiver and answered, "Yard office."

"Is the division foreman around there?" asked the caller.

"Hang on," returned Washneck as he placed the earpiece on the desk. Leaning out the doorway he shouted, "Mr. Langland, you're wanted on the phone."

"Langland speaking," he answered.

"Jack Rice here in Green Bay," replied the superintendent. "The dispatcher is in a hurry to send out a work train. How soon can you have an engine ready?"

"The 475** is available, but there is no fire in her," answered Langland.

"Does she have a plow?" inquired Rice.

"Just finished putting one on," was the reply.

"Good," returned the superintendent, "Get her hot and notify the dispatcher when she's ready."

"I'll take care of it, Mr. . . . ." click. Before finishing the sentence, Rice had hung up. Grabbing his hat, Langland pulled on his Mackinaw. While moving past the clerk's desk he was busy buttoning-up. "I'm going over to the roundhouse

---

*During the winter months, the North Western replaces the cowcatchers with a low-profile snowplow on some engines.

**A 4-6-0 wheel arrangement.

to get someone to build a fire in the 475. Keep an eye on things," he added. On his way out of the office he nearly ran down the callboy.

After tossing his coat on the table, the callboy rubbed his hands together and said, "What's the big rush with Langland?" Washneck was beginning to explain when the phone sounded off again.

The clerk announced in his usual monotone, "Yard office." After a pause, during which he was obviously receiving some instructions, he interrupted long enough to say, "Hold on a minute till I line up a pad to copy this down." With a pencil locked between his one finger and thumb and the receiver wedged between his ear and shoulder, Washneck continued, "OK, I'm ready, shoot." Hastily, the clerk began taking down the message.

He held the paper so as to take advantage of the single overhead light bulb, and began to read it back. "Switching crew at North Fond du Lac to make up a work train consisting of the following: One tool car, one empty boxcar, three outfit cars,* and one coach. This train to be ready on the house track by 11:30 tonight. Is that it?" asked the clerk. "OK Chief," added Washneck, "I'll get it to them right away." Washneck held the message out to the callboy. "This one's hot, get it to the head switchman out in the yard."

"Of all the nights, why does he pick one like this to run messages," grumbled the callboy as he wrapped up and jammed his shoulder against the door.

✿ ✿ ✿

The tall drive wheels on the Atlantics were designed for speed. When it came to plowing snow, they were at a definite disadvantage. Strange enough, however, the faster they went the stronger they became. Consequently, the way to overcome this weakness was to keep the speed up. Often this meant working the engine with wide open throttle which used up a lot of steam and rarely gave the fireman a chance to rest.

Between Rosendale and Ripon, there was a stretch of marshland. The ground was level and the snow barely covered

---

*Outfit cars were boxcars rebuilt into bunk cars for the track workers.

the rails. The engine responded by gaining speed. Soon the swaying and pitching became quite pronounced and several times the fireman had to check his swing in order to keep from missing the opening. Hoping to pick up a little time, the engineer was reluctant to ease off on the throttle.

Sharply outlined against the white background, the beam of the headlight picked up the "one mile to station" whistling post. Hassman reached overhead and held the whistle open for one long blast.

Conductor MacDonald was counting tickets when the sound of the whistle penetrated the coach. Instinctively he pulled out his timepiece. Sitting across the aisle facing him, the trainmaster followed suit. "We are nearly 40 minutes late," commented the conductor.

"I felt her slow down several times, must have hit some good sized drifts," replied the trainmaster as he wound his watch and returned it to his vest pocket.

MacDonald picked up his lantern and headed for the rear of the coach. "Ripon-Ripon," he announced, lightly touching the backs of each seat on the way.

"Everybody out this way," shouted the conductor as he opened the vestibule door and proceeded to raise the platform over the steps.

❀   ❀   ❀

Looking ahead along the boiler, Hassman recognized the station agent standing on the far end of the platform with his back to the wind.

As soon as the engine came to a stop, Hassman picked up the oil can and started climbing down. A high-pitched voice behind him said, "The dispatcher wants you to give him a report on snow conditions."

In order to complete the oiling before the conductor gave the highball Hassman knew he'd have to hurry. Moving over toward the drive wheels he motioned with his head and said, "Follow me." With earmuffs attached to his cap, collar turned up, and clipboard in hand, the agent patiently followed. The engineer poked the long spout through the spokes and turned toward the agent. "Tell the dispatcher we were bucking 8 to

12 inches of snow most of the way. Some of the cuts were over 2 feet deep."

With jerky movements the agent managed to copy down the message. Hassman started to cross over in front of the engine. "Is that all?" shouted the agent.

"Ain't that enough?" returned the engineer as he continued stepping over the rail. The agent quickly scurried back to the warmth of the depot.

Upon returning to the cab, Hassman glanced up at the steam gage, and said to his fireman, "I'm going to work this engine for all she's worth out of here; so get a good fire in her."

The fireman nodded assent and turned to look back. The passengers were all aboard and the conductor was giving the highball. "Let's go," shouted the fireman as he reached over and shut off the injector.

Hassman took his position at the throttle and started the train. Leaving town, the locomotive was cutting a swath 11 feet wide and 6 inches deep. It took over a mile to get the train rolling around 40 miles per hour. The whitened countryside added to the brightness of a full moon. Mist-like formations were skipping over the fields.

The fireman was on the deck heaving coal when the tangent* to the reverse curve came in view. "Get a good hold," shouted Hassman. Heading into the curve, the engine thrust its weight against the outer rail.

In less than two car lengths the snow was four feet deep. The engine felt as if some giant hand was suddenly pressed against the front end. Hassman and his fireman ducked behind the boiler as chunks of snow were disintegrating against the forward part of the cab.

The slack came in from the coaches and for a moment the added weight cancelled out the retarding effect of the snow. In a matter of seconds the engine came to a halt. When the air cleared, the front end of the smoke box was buried up to the headlight.

---

*Tangent track is that portion immediately adjacent to a curve.

Hassman leaned out of his side window and saw that the snow had caved in and covered the top of the 81-inch drive wheels. Looking back, he observed that the tank and coaches were in the same shape. He slammed the window closed and glanced over to the fireman. "Well, did you bring a deck of cards?" he asked half seriously. "Unless I miss my guess, we'll be here for a long time." Almost disdainfully he threw the Johnson bar over into the back-up position. After opening the sander he carefully eased the throttle out six or seven notches. Nothing happened. With both hands he continued pulling the throttle until she was nearly wide open. Suddenly the drivers broke loose and began to spin. Hassman shut off and with a disgusted expression on his face he announced, "We're stuck." He grabbed the whistle cord and called out a flag.*

"How far are we from the depot?" inquired the fireman as he anchored the firebox door open.

"I should judge about two miles," replied Hassman.

The canvas curtain was pulled aside and the head brakeman leaped down onto the deck. Having walked through the cars he arrived by climbing over the tank. He moved over in front of the open firebox, pulled off his gloves and started rubbing his hands together. Looking over toward the engineer he said, "What's the matter Al, are we going to have to fit this baby with snow shoes?"

Hassman crossed his arms over his chest, leaned back, and remarked, "I'm open to any suggestions."

"All right," returned the brakeman, "here's one from the trainmaster. The rear brakie is on his way back to the depot in order to get instructions from the dispatcher. In the meantime the trainmaster wants you guys just to sit tight." As he started to leave he added, "I got four stoves** back there going full blast and at the rate they are burning up the fuel, we'll be out of coal in another four hours." With that he pulled the curtain back, climbed over the coal gate, and started back.

          ✿   ✿   ✿

*Sent the rear brakeman out with his lantern to protect the hind end from possible collision.

**The wooden coaches were heated by potbellied stoves.

144

During the long walk back to the station it was the heat from his kerosene lantern that kept the brakeman from nearly freezing to death. When he entered the waiting room, it was deserted. In a corner of the room, obscured by a huge safe, the agent had leaned back and crossed his feet in the seat of an adjacent chair, evidently taking a little nap.

Still suffering the painful effects of the bitter cold, the brakie was a bit unhappy at the sight of the agent enjoying solid comfort. "Is anybody here?" he shouted at the top of his voice.

The agent woke with such a start that he nearly fell back against the water cooler. Fumbling nervously for his pipe, he finally regained his composure, and said in a subdued voice, "Looks like No. 9 didn't make it through the cut."

"You're so right," answered the brakeman, "and we ain't gonna make it back here either unless we get help right away."

"Come on in the office." As the agent spoke, the pipe in his mouth bobbed up and down.

"Get the chief dispatcher on the phone," demanded the brakeman.

The company phones on the branch lines consisted of wooden boxes with adjustable mouthpieces and were mounted on the wall. A crank on the side provided the power for a signal system.

The agent was busy searching through his vest pockets for a match. Giving up in disgust, he put his pipe on the table and stepped up to the mouthpiece. After several turns of the crank, he paused to listen. "Hello, dispatcher? This is the agent at Ripon, is the chief around? Tell him No. 9 is in trouble and the brakeman wants to talk to him." Handing the receiver over, he moved back and continued searching for a match.

"Hello, chief? We're hopelessly stuck in six feet of snow. Two miles north of Ripon. MacDonald, my conductor wants some instructions." After a pause, he answered, "OK." Turning to the agent he said, "Here, the dispatcher wants you to take down some orders."

The agent took out some carbons and slipped them in between the flimsies. Placing the pad on the little platform directly below the phone, he grabbed a pencil. "All right, chief, I'm ready," said the agent.

As he copied down the instructions, the agent repeated them aloud. "Extra west due to arrive 2 miles north of Ripon at approximately 4 a.m. Crews on No. 9 and work extra shall establish contact soon as possible. Roadmaster Del Lorenzo will supervise the removal of snow. Trainmaster Thompson shall take charge of all train movements. Every effort must be made to ensure the safety and comfort of the passengers. Agent at Ripon shall arrange for lodging of travelers held up enroute. A joint progress report shall be wired to the division superintendent and chief dispatcher every hour on the hour. All scheduled trains between Fond du Lac and Wisconsin Rapids are hereby cancelled until further notice."

When the message was completed, the brakeman spoke up. "Tell him to hold on." The agent handed the receiver to the brakeman and stepped back. "Say Chief," began the brakeman as he leaned over the mouthpiece, "we have nearly a hundred passengers out there and barely enough fuel to keep those stoves going another three hours. Even if that rescue train arrives on time we will have been out of heat for two hours." During the chief's response the brakeman nodded his head several times in apparent agreement. At the conclusion he replied, "Okay chief, that's just what we'll do." After replacing the receiver on the hook he commented half aloud, "Now why didn't I think of that?"

"What did he say?" asked the agent.

Sheepishly, the brakie replied, "Put all the coal in one car and transfer the passengers into it. We can always rob the coal car on the engine if that runs out."

"Smart guy that dispatcher," commented the agent as he removed the orders from the pad. Handing a copy over to the brakeman he added, "I guess your conductor will be anxious for this."

While the brakeman started to pull on his gloves, the agent poured out a cup of steaming hot coffee from a thermos. "Here, this will help keep you warm on the way back," he said with a smile.

The brakeman thanked him, picked up his lantern and started back to the train. As he trudged along in the blowing snow, he realized that the agent's coffee seemed to generate warmth in his heart as well as his body.

13.

To the Rescue

Central Wisconsin shared the prosperous years preceding the depression and during this period growth was phenomenal. Two switch runs* were required to gather the freight from the surrounding territory and deliver it to Wisconsin Rapids.

There were three large paper mills; one at Wisconsin Rapids, one at Port Edwards, and the other at Nekoosa. The manufacture of paper utilizes a lot of pulpwood. In order to deliver this material, one train would make a trip north to Marshfield with a string of empties and return the same day loaded with pulpwood. The following day, the other switcher would distribute these loads at Port Edwards and Nekoosa. On their way back they carried the paper along with the empties.

Nos. 37 and 38 were time freights assigned to handle the business between Wisconsin Rapids and Fond du Lac. This arrangement provided service each way every day.

The crews on the switch runs were from Fond du Lac. Working away from home, the men could be with their families only once a week. On Saturdays it took real planning and a bit of luck to enable them to catch the passenger for Fond du Lac.

A small hotel, converted from a spacious old mansion, provided lodging. It was operated by an elderly couple and meals were home-cooked and served family style.

Both jobs required 12 to 14 hours on duty, leaving very little time for anything beyond eating and sleeping. The close

---

*Switch runs were a combination switcher and freight train.

association for prolonged periods gave the men an opportunity to become well acquainted. Each one knew what to expect from the other with the result that the job was an outstanding example of teamwork.

Swain Rasmussen, the conductor, hired out in 1905. Known as the "Determined Dane," he had a reputation for getting the job done. With a bald head and a generous midsection, he reminded one of an oversized version of Disney's Grumpy. He possessed an enormous appetite, and always made sure his caboose was equipped to prepare meals enroute. Besides being an excellent cook, he was a generous fellow, a combination which made him popular with the crew. His menu varied from sauerkraut and pig's knuckles, to sauerkraut and spare ribs to sauerkraut and wieners. On occasion he would treat the boys to a good old fashioned stew. Needless to say, none of his co-workers ever suffered from constipation.

Then there were brakemen Lauderman and Moss, both energetic young fellows, eager to perform their duty and go beyond if necessary.

Clarence "Dutch" Blader was the fireman. Having faithfully served his country during World War I, he was intensely loyal. Not a fighter by nature, he would nevertheless tolerate no reflection on his beloved flag. He was a likeable personality with a ready wit. His presence on the engine helped to lighten the long weary hours.

During the colder months many of the older engineers would avoid the more strenuous jobs. Consequently, by the time winter set in, their runs would be up for bulletin.* Being one of the younger engineers and having possessed an abundance of health, George never hesitated to accept any assignment. When the Wisconsin Rapids-Marshfield switch run was up for bid, George caught it. Within a week he was sharing sauerkraut and fellowship with the "Determined Dane."

❋   ❋   ❋

Both crews crowded into the ticket office to pick up their orders for the day. The agent's hand was full of papers as he

---

*Open for bid on the basis of seniority.

turned in his swivel chair to greet them. "Both switch runs have been cancelled for today," announced the agent.

"That's a fine how-do-you-do," complained Rasmussen. "How come?" "Don't get excited," returned the agent. "It just so happens that the chief dispatcher has selected you boys to punch a hole through to Fond du Lac." The announcement was received with mixed emotions. An extra day at home was most enticing but the magnitude of the task and the dangers involved were well understood.

Louis Gill, the engineer on the other switch run, stroked his chin thoughtfully and stared out the office window. After a pause, he looked down at the agent and remarked, "Can't understand why he called on us to do this nasty job." "Wouldn't surprise me if Ole Granny broke down before we even got to the deep snow," added Rasmussen. "Seems to me," said Blader, "this should be a job for one of those rotary rigs."

The agent got out of his chair, handed a copy of the orders to Rasmussen and said, "While you guys are griping, remember, I don't give the orders, I just hand them out."

George pulled out his watch and did a little mental arithmetic. "Time's awasting, fellows, give Rasmussen the floor so he can read us the orders."

The conductor adjusted his glasses and leaned toward the window so as to take advantage of the morning light. "Effective this a.m. both switch runs are cancelled. A work train consisting of one tool car, one coach, a caboose and plow number 28 [Ole Granny] shall be doubleheaded with engines number 1402 and 1078. Proceed to Fond du Lac at once. Be advised snow depth in the reverse curve north of Ripon exceeds 15 feet. Engineer George Williams and conductor Swain Rasmussen will be in charge of the train."

When the conductor finished, he looked over the top of his glasses and said, "Well, Buddy, what do you make of it?" Glancing outside, George saw that a light snow was beginning to fall. "Let's go over to the roundhouse and have a look at that plow," he replied. "In the meantime, you boys get the engines hot,"* he directed both firemen. On the way out the door, Rasmussen told Moss to get a fire started in the caboose. It was just a short distance, but the snow was knee deep and by the time they got there, both Rasmussen and Gill were puffing. Snowplows would rarely see more than two weeks of actual service in any one year. Many of them remained inactive two or three years in a row. Nonetheless, when the need arose it was always an emergency. In order to minimize the cost for such equipment the North Western designed a heavy iron blade which could be fitted to the end of any standard gondola. Loaded with a hundred thousand pounds of rock, the gondola proved to be a stable platform.

Unlike the rotary, the flying wedge type plow depended upon brute force to knife through the drifts and clear a path. With a good engine they could handle three or four feet of snow on straight and level track.

Whenever the snow became so deep as to stall the engine, the hoghead would size up the situation and determine if it was practical to back up and take a run for it. Experienced engineers knew just about how much of a run it took to penetrate a given amount of snow. In cases where the depth continued for an extended distance, the hogger would back to the nearest telegrapher and call for another engine.

Plowing on curves was considerably more dangerous. By the time the front wheels of the gondola began to follow the curve, the plow had penetrated straight ahead for nearly fifteen feet. As the body of the gondola continues to turn, the blade is swung sideways. The sweeping action exerts greater pressure on the side of the plow which is angled toward the inside of the curve. This force can result in a derailment. Here is where the load within the gondola plays an important part. The added weight helps keep the flanges following the rail.

Unfortunately, curves were often cut through hills and these troughs formed perfect traps to collect the snow.

The speed required for a breakthrough depended upon the power available (size and number of engines) and the weight of the entire train. The degree of curve, the condition of the

*Build up the fire in order to produce a working steam pressure.

151

track, the amount of elevation (bank) and many other factors had to be considered. In the last analysis, it was the engineer who shouldered the responsibility for his own decision.

A heavy white mantle gave Ole Granny a ghost-like appearance as it stood alongside the roundhouse.

"You fellows wait here, I'll be right back," said George as he hurried toward the roundhouse. A short distance inside the door, he found the foreman busy at a work bench. "Can you board up the front windows on both our cabs?"* he asked. "What's the occasion?" returned the foreman. "Instead of going out on our regular run, the dispatcher has assigned us the job of clearing the road to Fond du Lac." Looking over at the front end of both engines, the foreman shook his head and remarked, "I'd say he's expecting an awful lot of those two R-1's. No doubt you'll be shoving Ole Granny all the way," he added. "Speaking of Ole Granny, what condition is she in?" "Not too bad, but let me warn ya. With the iron plow and that load of rock, those journals might start running hot." "I'll keep an eye on her," assured George. "By the way, can I borrow a hammer?" Pulling open a bulky drawer, the foreman tossed one on the bench. Snatching it up in one hand, George grabbed a broom with the other and hurried back to the plow.

George hastily swept the snow off the areas which anchored the plow to the gondola. Exchanging the broom for the hammer, he proceeded to tap each bolt. Any movement to these fasteners would indicate a critical weak spot. In spite of her age, the old gal seemed solid.

Raising the cover of one of the journals,** George saw that the packing looked good and the oil was fresh. Moving over to the front end, he looked up at the heavy iron blade. She was taller than most plows and the contour of the broad wings was designed to throw the snow high and wide.

While reaching over to take the broom, George observed that both men were stomping their feet and showing signs of

---

*Snow thrown up from the plow frequently caves in the front window of the cab, sometimes resulting in injury to the enginemen.
**External bearing area on the end of the axle.

becoming impatient. "You boys getting cold?" he asked. With chattering teeth and quivering voice, Rasmussen quipped, "What gave you the clue?" "Let's get inside," said George as he headed toward the roundhouse. The two men hobbled stiffly along behind.

After stashing the broom back of the door, George tossed the hammer on a nearby bench. It was a small roundhouse and the noise from the blowers on both engines made conversation difficult.

Engineer Gill unbuttoned his coat, shook off the loose snow and raised his voice. "Well, Buddy, do you think Ole Granny can do the job?" George was squatting down inspecting the flanges on the pony tracks (lead wheels). Looking up he answered, "I'd like to hook onto her and make a running brake test before I decide." George looked up and saw the foreman nailing the boards in place.

Rasmussen turned toward Gill. "Why don't you and your crew go ahead and make up the train while Buddy and I take Ole Granny out for a test run?" Gill nodded his head and moved off toward his engine in the adjacent stall.

❖　　❖　　❖

Both of these locomotives were called "Ten Wheelers." This meant that there were four wheels on the pony tracks (lead wheels) and six drivers. Though somewhat bigger, they were very similar to the famous No. 382 on which Casey Jones made his fateful last ride.

This wheel arrangement gave the engine good traction and it also made it possible to negotiate the tight curves found on branch lines.

The North Western gave this engine the designation "R-1." Originally intended as a dual purpose locomotive for use on the main line, it faithfully served in this capacity for over 25 years. The R-1 proved to be one of the most reliable engines in the North Western's stables. However, like many of the designs developed before the turn of the century, her power became inadequate for the heavier trains. With an excellent record for durability, the R-1 was selected to fill the growing demand for power on the branch lines.

Through a series of improvements, the steam locomotive evolved into an efficient mechanical giant. Anything that failed to contribute to its economic function was ruthlessly discarded. Individual names which once graced the side of every cab gave way to numbers. The polished brass housing of the oil lamp with its fancy decorative scrollwork was replaced with a small electric bulb inside a black box.

The manufacturers of the R-1 managed to retain a few of the refinements which personalized the earlier locomotives. A contoured leather grip, much like that found on a fine hunting knife, formed the handhold on the throttle. This control moved from the shut off position to wide open with a minimum of effort. The action was free from tight spots so frequently experienced on the bigger engines. The forward window on the cab was wide enough to provide a broad view of the track ahead.

*　　*　　*

Rasmussen followed George up into the cab. Blader finished dropping a couple of scoops of coal into the back corners. The steam pressure had climbed to 80 pounds. George opened the valve to the air pump and started checking his gage cocks. The fireman slid the scoop under the coal pile, climbed up on his seat box and adjusted the blower.

George was about to start filling the lubricator when Blader spoke up. "I've got her filled to the top."

"Thanks much," returned George as he proceeded to adjust the rate of oil flow.

Placing the monkey wrench in his hind pocket, George raised the long spout out of the rack and started down off the engine. The fireman grabbed the broom and began to sweep up a few chunks of coal. While George was adding a couple of turns to the grease plugs on the main rod, the doors back of the other engine swung open and engineer Gill backed her out on the turntable. Due to the position of the valve gear, George couldn't reach one of the bearings. Moving around the front of the engine to the fireman's side, he hollered, "Hey, Dutch, throw her over in the back-up position!" The fireman first made sure the throttle was completely shut off, and then threw the Johnson bar over. George went back and aimed a good squirt at the proper location.

By the time he returned to the cab, the steam was up to 125 pounds and the air gage was registering 90. "Dutch, will you ask the roundhouse foreman for two buckets? I'd like to carry some extra sand in case we run out." Looking over to Rasmussen he added, "I'll be all set when he returns." The conductor moved over to George's side and commented, "Buddy, we have been in a few tight spots before. This time I think the dispatcher bit off more than we can chew." "We're going to have our hands full all right," replied George, "but I've got a plan that I think will get us through." "Would it be asking too much to let me in on your secret?" Rasmussen asked. George removed one of his gloves and patted Rasmussen on his bald pate. "You've been a pretty good boy. Guess maybe I ought to tell ya," he said with a mischievous smile. Dutch Blader climbed up with two large pails and stowed them on the tank above the locker.

George draped his gloves over the brake valves. As he rested his foot on the reverse lever quadrant, he looked down at the conductor. "Did you notice those two hoppers loaded with coal just across from the depot?" he asked. "Yeh," replied Rasmussen. "They're headed for the mill at Nekoosa, so what about them?" "Well, you know they'll never get delivered until we return from Fond du Lac," said George. "What's that got to do with it?" returned Rasmussen.

After shifting the toothpick to the other corner of his mouth, George began. "Before we get to the deep stuff, we will have plowed snow for fifty miles. I figure the cut will average ten feet deep for nearly a quarter of a mile." The fireman hopped down on the deck and tossed in a layer of coal. As he slid the scoop back under the pile, he said, "Sounds to me like we're going to wind up freezing to death in the middle of nowhere." Climbing back up on the seat box, he adjusted the blower and added, "Frankly, I think the dispatcher has lost his marbles."

"Before you condemn the whole operation give Buddy a chance to explain his plan," returned Rasmussen. "I admit," said George, "we'll be taking a chance, but the situation is critical and a lot of our friends are counting on us."

Moss, the head brakie, climbed up into the cab. "You boys about ready?" he asked hopefully. "Buddy is outlining his plan to get us through," said the conductor. Moving over to the seat box alongside of Blader, Moss dangled his feet over the side and said, "Okay Buddy let's have it."

Glancing toward the steam gage, George noted that the needle had moved up to 150. Rasmussen began nodding his head impatiently as if to say, "I'm waiting on ya." George began slowly. As he spoke he pointed out that both engines would have to be pushed to the limit. At that rate they would be out of coal and water before they reached the reverse curve north of Ripon. Ordinarily, trains could pick up coal and water several places along this route, but with conditions as they were it would be foolhardy to depend upon them.

Glancing over the coal gate, Blader eyed the huge pile of coal. Looking back at George with one eyebrow raised he asked, "Are you serious about me moving that whole load into the firebox before we reach Ripon?"

"Don't panic, Dutch," replied George. "If it gets too much for ya, I'll take the scoop and you can run the engine."

Continuing, George suggested that they put one hopper "full of coal" back of each engine and let the section hands transfer the coal into the tanks as necessary. "Sounds like it might work, but what about the water?" asked Blader. "If we get low we'll stop and put the section crew shoveling snow into the tanks," replied George.

For a brief moment all four men were silent as they weighed the proposition. Finally, Rasmussen spoke up. "Buddy, I'm with ya all the way." Turning to the others he added, "Let's give it a whirl." Moss nodded his head. "Count me in too," said Blader as he eased off on the blower.

❀   ❀   ❀

The pop valves began to sizzle. "Let's back her up," said George as he opened the cylinder cocks and moved the Johnson bar over into the backup motion. Blader started the bell and soon the big doors back of the tank swung open. While

156

easing the engine on the turntable, Moss stepped over to George's side and inquired. "What's our next move?" "We'll fill up on water and sand and then go pick up Ole Granny," replied George.

Blader topped off the tank,* while the roundhouse helper filled the buckets with sand and loaded the sand dome. After moving the engine past the lead to the plow, Moss jumped off, threw the switch and signaled to head in. While closing the gap, the sound of the exhaust from the other engine indicated that they were busy making up the train.

When the engine coupled on, a solid jar was felt in the cab. The brakie connected the air hose, opened the angle cock** and gave a backup signal. It took a shot of sand and over half a throttle to start. George kept the speed down to a crawl until they pulled past the turnout onto the roundhouse lead. After he closed the switch, Moss ran to hop on.

The engine responded sluggishly as she sensed the heavy weight of the plow. Shutting off the blower, Blader looked over toward George. "That baby must weigh close to 90 tons," he remarked. "In weather this cold, the oil on those journals is about as thick as molasses. Once the bearings warm up she'll pull easier," explained George.

Crossing over onto the main, George noticed that the other crew had finished making up the train and Gill was backing it on the side track. Rasmussen moved over to the engineer's side and said, "While you're running the brake test, I'll get Gill to switch those hoppers into position." Nodding in agreement, George eased off. When the front of the plow cleared, he set the brake. Moss realigned the switch and gave a highball. As George reached for the throttle, he saw the agent waving at him from the platform. Sliding his side window open, he leaned out and shouted, "What is it?" "The chief wants to talk to you," returned the agent. "Tell him I'll call him back in 20 minutes," replied George.

---

*Filled it full of water.

**Valve controlling the admission of the trainline air pressure.

The brakeman gave George a high sign and said, "Turn 'er loose Buddy, let's see how she handles." After closing his window George moved the Johnson bar to the forward corner and released the air. As they moved past the depot, the section crew was busy loading shovels and equipment into the tool car.

Snow was a foot deep over the rails. George decided to clear a path leaving town and make the test on the way back. About a mile out, he brought the train to a gradual stop. After reversing the engine, he opened the throttle and started back.

In order to be sure Ole Granny was braking properly, George planned to get moving around 30 mph and then try a 20 pound application. By holding the engine brake in full release, the resulting action would provide a good indication of her braking capabilities.

A half mile from the depot, George held the whistle open · for one long blast. The speed had picked up to around 35. Holding the engine brake off he set the train's brake until the gage registered a 20 pound reduction. Lapping the valve, he waited for them to take hold. Soon the engine felt a strong pull. Looking over toward his fireman, George shouted, "She's okay." As they approached the depot he released all but five pounds and made a smooth stop right in front of the ticket office. "I see they are all ready for us," remarked Moss. Looking over through the fireman's side window, George could see the other engine had one hopper on behind and one in front.

George set the Johnson bar to top center and moved off the seat box. "Dutch, will you couple into the train while I go in and see what the dispatcher wants?" Blader nodded, stepped over to the engineer's side and started the engine.

Before George made it to the depot, the agent started ringing for the dispatcher. Lauderman and Rasmussen were standing in the ticket office as he came through the door. "How'd she go?" inquired the conductor. "The brakes are working good," replied George. Stepping over to the phone, he took the receiver from the agent's outstretched hand. "Hello Chief, what can I do for you?"

"Let me explain the situation," said the chief. "It took a work train with 50 section men four hours to get No. 9 back to Ripon. We have nearly a hundred passengers stranded at the local hotel and the company is picking up the tab. I sent out no less than six engines from Fond du Lac. Four of them are stalled on the siding at Eldorado and two more are dead at Rosendale. The paper companies are screaming for deliveries. I tell ya Buddy my back's against the wall." "Hold it chief, just give me the punch line, the whole train crew is waiting on me." "I know I am asking for a miracle," continued the chief, "but you're the only hope I have left. Do you think you can make it?"

"You understand chief," replied George, "my front windows are boarded up and I'll be running practically blind. Can you assure me that there is no rolling stock blocking the line?" After a short hesitation he answered, "Everything between Fond du Lac and Wisconsin Rapids is in the clear." George glanced at his watch. "You can tell Al Hassman on No. 9 to get his engine ready 'cause I'm going to be busting through there around 5 p.m." "Okay Buddy, I'll see to it that they are ready to go."

As George hung up the receiver, Rasmussen slapped him on the shoulder and said, "By gosh, I believe you mean it." "Listen Rasmussen," returned George, "I'm putting that plow through to Fond du Lac if I have only one wheel left on that engine." The conductor started to button his coat. "That's the kind of talk I like to hear," he said as he started for the door.

Following along behind, George began to pull on his gloves. "Are we all set?" he asked. "Ready to go," returned the conductor. "Good luck," shouted the agent as the men left the depot. Lauderman and Rasmussen turned toward the caboose. They hadn't gone a hundred feet when George hollered, "Hey, Swain! Keep everyone anchored to their seats. I'll try to warn you with the whistle, but there is no telling when we might run into a solid drift." Rasmussen nodded his head to indicate that he understood. Walking over to the second engine, George stood directly below the cab. Gill opened

159

his window, looked down and remarked, "What's the story, Buddy?" "Section crews will be getting on board at Bancroft and Wautoma. Conditions permitting, we'll take on coal and water at the same time," replied George.

"Where do you want to tie up* for dinner?" asked Gill. Pulling out his watch, George studied it for a moment. "Let's try to make Green Lake," he answered. "I'm ready to go whenever you are," he returned. Responding with a wave of his hand, George hurried back through the snow. Pulling the canvas curtain aside, George took his position at the throttle. After a quick check of the gages he looked over at the fireman. "All set?" he asked. Blader gave him a high sign. George slid his window open, leaned out and looked back.

Standing on the bottom step of the caboose, Rasmussen was giving out with a highball. Answering with a couple of tugs on the whistle cord, George started the train. After all the wheels were moving, he eased off four or five notches and held the speed to a fast walk. About the time the caboose progressed to the crossover, the fireman looked back to be sure Lauderman had time to close the switch and get on board. When the rear brakeman reached for the grab iron, he gave a highball. "We got 'em," shouted Blader as he reached for the scoop and started stoking the fire.

Twin columns of black smoke belched skyward as both engineers widened on the throttle. The bark of the two locomotives could be heard all over town. By the time they emerged into the country, the train was rolling 30 mph.

The roundhouse foreman had been considerate enough to leave a half-inch gap between the boards which covered the front window. By leaning forward, George could barely make out the front end of the plow.

It was flat terrain all the way to Bancroft (15 miles) and the snow offered very little resistance. When the train reached around 40 mph, George eased off to a half-throttle. By observing familiar landmarks along the route, George had a pretty good idea where he was at all times. Frequently, he

*Refers to stopping the train.

would pull down on the bill of his cap, open the side window and face into the wind. A small deflector, just in front of the arm rest, helped to shunt away the icy blast. However, a light snow was falling and in a matter of seconds his face would become completely covered. Each time he leaned out, it was a painful experience.

Coming into Bancroft, George shut off to a drifting position, held the whistle open and simultaneously made a light brake application. (Before the brakeshoes take hold properly, they must be warmed up.) In about 30 seconds the slack began to stretch out.

Some of the buildings on the outskirts of town went by. George leaned out to gage the remaining distance to the depot. By adding another 10 pounds of air in the brake cylinders, he brought the train to a stop directly under the spout.

Blader climbed over the tank, and took water. George moved the train ahead until the tank on the second engine lined up with the spout. When the fireman pushed the spout up into the clear, George moved the train on down to the coal shed. Blader dumped the chute until the coal began to fall over the sides. After spotting the coal car of the rear engine under the chute, the rear fireman repeated the process. While this was going on, the section foreman climbed up into the cab and greeted George. "Me and my boys are planning on coming back on No. 9 tonight. Now don't you let us down." "I'll try to punch that hole through on the first run," returned George. "But if we do get stalled get your men out on the double. The snow must be cleared away from the plow and out from under the train before we can back up and take another run for it." "You can depend on us," said the section foreman as he started off the engine. Leaning out the window, George hollered, "Keep your men glued to the seats. We might hit pretty hard and the jar could hurt someone."

Soon Rasmussen gave a highball and the train was leaving the depot. About two miles out of town the flat country began to change to a series of low rolling hills. In some places the snow was over three feet deep. Both engines were wide open, but the speed dropped below 20 mph several times.

Blader was so busy heaving coal, he had precious little time to rest. As the town of Almond faded in the distance, George took the scoop and told Blader to run the engine. Actually, George enjoyed an occasional opportunity to fire. Men who watched him handle the scoop were always impressed at the apparent ease with which he fired the most difficult jobs.

The farming community of Wild Rose slipped on by. Wautoma was six miles ahead. Before entering the city, Blader called over to George. "You take her on in. Thanks for the rest." Approaching the depot, the train order board was clear. During the last 25 miles, both engines had used a lot of fuel and water. Fortunately, the coal shed and water tank at Wautoma were in working order.

While waiting for the second engine to take on coal, George climbed off and trotted to the front end of the plow. Removing one of his gloves, he held his hand directly above each of the journal boxes. The wheel bearings on the front end were radiating a little heat. But it was nothing to be alarmed about. Before leaving Green Lake he planned to squirt a generous amount of oil on each journal. Hurrying back toward the engine, he observed 10 or 12 section hands lined up to get on board. Back up on his seat box, he leaned out and looked back.

Conductor Rasmussen waited for the fireman to stow the coal chute and climb back in the cab before giving the highball. With a couple of toots, George started the train.

The train order board was clear at the little station of Neshkora. The agent was out on the platform giving a hearty wave as they rumbled on by. Another eighteen miles and they would be pulling into Green Lake, where George figured they should arrive by one o'clock. Looking over toward Moss and Blader, he shouted, "Getting hungry?" Blader gave him a toothy grin and hollered back, "I could eat a horse — hoofs and all."

Conditions were getting worse. Several times drifts rose over four feet. As the resistance built up, George felt the weight of the train come in and literally push him through the bad spots.

It was beginning to appear as if the extra coal in the hoppers would not be needed after all. But without saying so, George had a far more important reason for hauling them along. The weight of those two loaded hoppers might just provide the additional weight necessary to blast their way through.

Princeton was about midway between Wautoma and Green Lake. Heading south, the depot was on the engineer's side. The train order board was clear. While passing the ticket office, the agent could be seen busy at the key. No doubt the dispatcher was getting a report on their progress.

The roadbed was quite level all the way to Green Lake and being slightly elevated above the ground, the snow barely covered the rails. George had been hoping for an opportunity to observe the way Ole Granny handled at higher speeds. A veteran at plowing snow, previous experience had taught him that speed was a prime factor to success.

The North Western used light track (72-pound rail) on their branch lines. As a result, irregularities in the joints were more noticeable.

George pulled the throttle about ⅔ open and positioned the Johnson bar two notches ahead of dead center. As the engine gained speed, the rough track accentuated the pitch and roll. Leaning out to determine how the plow was riding, George was encouraged to note that she was taking it smoother than the engine.

Soon, several landmarks indicated they were nearing the outskirts of Green Lake. After giving one long blast, George started the braking action. As expected, the train order board, on top of the depot, was out.*

He stopped the engine on the far side of the platform, centered the reverse lever and set the engine brake. While stepping down from the seat box, he motioned with his thumb toward his mouth. The two men followed him off the engine.

---

*Indicating the agent had orders from the dispatcher.

Both crews hurried over to a quick-lunch counter across from the depot.

As the boys sat down, Rasmussen suggested, "Better get a good feed into ya; no telling when you'll eat again."

Pointing at the conductor's big stomach, George commented, "With a pouch like that, you could go without eating for a week and still have plenty in reserve." The boys enjoyed the jest at Rasmussen's expense.

The food was excellent and the hot coffee hit the spot. While lining up at the cash register, Rasmussen patted his full belly and said, "You boys go ahead and make fun of me if you want to, but don't forget that I got in this shape by following the doctor's instructions." "Get a load of this, fellas," announced George. "Swain is blaming his doc for that big breadbasket." "You can't be serious," remarked Blader. "That's right," continued Rasmussen, "after my last examination, the company physician told me I should be careful and keep a close watch on my waistline. Since then, I've been getting it out where I can see it." The Determined Dane joined the boys in a hearty laugh.

Placing a toothpick in the corner of his mouth, George paid for his dinner and headed for the depot. When both crews were gathered in the ticket office, the agent handed a copy of the orders to Rasmussen and George. Moving over alongside of Gill, George held out the flimsies and they studied them together. They read as follows: "Snow depth averages three feet for 5½ miles south of Green Lake. One hundred feet north of the reverse curve, the depth increases until it reaches the maximum at the tangent between the two curves. From this point all the way to Ripon, the snow averages about two feet. Caution: Due to a hard frozen crust on the surface, all personnel are warned to take cover during the remaining operation.

"Engineer Williams is advised to continue past the switch beyond the north end of the depot at Ripon. This will permit No. 9 to back out on the main line. No. 9 will not leave until section hands have sufficient time to transfer."

"From the way this order reads, I'd say the dispatcher has a lot of confidence in us," commented Rasmussen. Folding the flimsies and slipping them into his pocket he added, "Shall we get going?"

Turning to Gill, George said, "Bucking three feet of snow before we even reach the bad spot is bound to limit our speed." "Well, what do you intend to do?" asked Rasmussen, sounding a bit impatient. "Under the circumstances, I suggest we plan on making it in two passes," replied George. "The first one can be at moderate speed with the intention of clearing a path as far as we can go. Then while the section hands are digging us free, I'll get out and look the situation over. We'll back here to Green Lake, replenish our coal and water, and make our second run at high speed," he added. "Sounds all right to me," commented Gill.

"I'll get the guys anchored down in the coach. Whistle off when you're ready and take out," said Rasmussen as he started back for the caboose.

Turning toward Gill's fireman George said, "I carried two extra pails of sand. Empty one in your sand dome. Blader will empty the other in ours. Might be a good idea if both you boys opened the sanders and tapped the delivery pipe in front of the drivers. Sometimes they get plugged with snow."

George climbed up into the cab and reached for the oil can. Moss took his place on the fireman's seat box. "What are you doing up here?" asked George. "I just want to be where the action is," replied Moss. "Now you listen to me," returned George. "There are no switches to throw between here and Ripon and we have no use for you on this engine." "Aw, come on Buddy, I want to see the grand slam." Moss was not aware of the danger involved. With his patience running thin George responded in a firm voice, "Sorry, Moss, I'm not asking you, I'm telling you to get off." Moss hastened to leave.

While Blader was busy filling the sand dome, George started making the rounds with the long spout. Before heading back he gave each journal on Ole Granny several good squirts. He saw that Gill also was applying the finishing touches with his oil can.

Blader turned up the blower and started stoking his fire. George stomped off some snow that was clinging to his overalls. He slid his windows open and looked back. Gill stuck his head out and gave a high sign.

"Here we go," shouted George as he reached overhead, made two short blasts and started the train. Leaving town, the snow was about six inches over the rail. As soon as they hit open country the level increased to about three feet. With that much resistance, both engines labored to gain speed. The crack of the exhaust emanating from behind was evidence that Gill was getting all he could out of his engine. Carefully he listened for the first sign of slipping. Each time she threatened to break loose, he gave her a shot of sand. Finally, at about 35 mph, it became apparent the train had reached its terminal velocity.

Soon the landmarks indicated they were nearing the reverse curve. George leaned way out. Less than a quarter of a mile ahead he could see the entrance to the cut. Blader was on the deck heaving coal. While reaching overhead for the whistle cord, George shouted, "Get yourself set, we're heading into the curve." With a series of short blasts, he warned the men riding behind.

Leaning out for one final look, George saw for the first time the huge mountain of snow. Slamming the window closed, he braced himself against the reverse lever. Surprisingly enough, it turned out to be a quick deceleration instead of a jolt. However, about the time George straightened up, he felt a sharp thrust as the weight of the train came in and pushed them forward into the deepening drift. The speed was down to 15 mph before George had time to advance the Johnson bar.

Looking ahead, through the gap in the boards, he could see huge chunks of snow tumbling over the top of the plow.

After struggling against overwhelming odds, the two engines grudgingly came to a halt. George no sooner shut off on the throttle when the pop valves let go. Quickly the fireman put on his injector and anchored the firebox door open. In a matter of seconds, the noisy column of escaping steam collapsed. Looking over toward George with a disgusted expression, he remarked, "Did you see that pile of snow coming over the top of the plow?" "I sure did," returned George. "It must be at least 20 feet over the rail."

The snow formed a vertical ledge the top of which was level with the arm rest. As he glanced down along the narrow gap, he felt relieved to find that very little snow had caved in around the wheels. "Let's try to back 'em up," said George, as he heaved the Johnson bar over. The fireman shut off his injector and closed the firebox door.

With three short blasts, George opened the throttle. Before the first exhaust cleared the stack, the second engine picked up her share of the load. However, both locomotives stalled just as the caboose began to move. Looking ahead, George could see the extended wings of the plow were wedged hard against the packed snow. Further effort under these conditions was useless.

George whistled out a flag (a signal to send the section men forward). Reaching out, he dug out a handful of snow. "This stuff is hard enough to walk on," commented George. "I'm going out and have a look at the situation up ahead." He crawled through his side window, and carefully stepped out onto the snow. The crust was so firm that his steps hardly left a print. At the front end, he could see where the plow had thrown the snow over fifty feet to the side and formed a continuous bank along the right of way.

The attempt to back up left a three foot gap between the plow and the snow. The front surface of the blade seemed none the worse for the wear. Continuing on around the curve, George stopped at a spot which appeared to be the highest point. He moved back a few feet and sighted along the top of his smoke stack. From this position he estimated the snow to be at least five feet over the top of his engine. The situation looked almost hopeless.

On his way back he saw a string of section hands moving up in single file, each with a shovel over his shoulder. George waited at the front of the plow and soon the men were gathered around him. One of the foremen spoke up. "Buddy, where do you want us to start?" "How many men do you have?" asked George. "Fifty-two," was the answer. "All right, put ten men to work on each side of that plow and widen the path back of the blade. I am going to need clearance for at least one hundred feet before we can back out of here."

Quickly moving clear of the assembly, George took a position about 25 feet ahead of Ole Granny. Pointing his finger in the direction of Ripon, he continued, "Now for the rest of your men, I must have a trench 6 feet deep. One that will follow the center of the rail all the way through the second curve."

"How wide do you want it?" asked one of the foremen. "Just as wide as you can make it," answered George. Glancing at his watch, he added, "You'll have about three hours. Soon as we get that plow loose, we'll back up to Green Lake, take on coal and water and start back. When you hear my whistle, get your men way back off the right of way. I'll be coming as fast as those engines can turn a wheel." Almost as an afterthought he added, "If that plow takes a notion to jump the track, we could have quite a mess." As George started back, there was a moment of silence as each man reflected on the dangers involved.

Soon the foremen could be heard shouting instructions to their men. George was about to climb back into the cab when he sighted the conductor laboriously making his way to the engine. Puffing hard as he arrived alongside of George, he was unable to speak until he finally caught his breath. Noting the snow level was way over the top of the plow, he complained, "The dispatcher's 15 feet looks more like 20 feet to me." "Could be he didn't want to discourage us before we got started," replied George. "About how long will it be before we can head back to Green Lake?" "Follow me," said George as he moved toward the front of the plow. Observing the progress they had made up to this point, George replied, "I'd say we'll be on our way in about 45 minutes."

Moss walked up with a shovel over his shoulder. "Where you going?" asked the conductor. "Buddy won't let me ride the engine, I'm no use to you back there, so I figured I might lend a hand with the shovel." "Good boy," said George as he patted him on the shoulder. "If we ever make it through, it will be the fellows with the shovel who made it possible."

Ahead he saw the section men lined up over the center of the track about eight feet apart and snow was flying in all directions. The sun was out and the air was still. George and Rasmussen walked back to the second engine. Gill opened his window and asked, "How're we doing, so far?" "Not too bad," answered George, "but we have a bigger task ahead of us than we figured."

Rasmussen took off toward his caboose and George started back to his engine. Just before returning to the cab he spotted another fifty or so men way down the track, each had a shovel on his shoulder. Apparently, these men were sent out from Ripon. The situation began to look a little more hopeful.

Back in the cab, George stood in front of the open firebox to thaw out. Comfortably perched on his seat box, Blader was leaning back with his feet propped up on the boiler head. The expression on his face gave George a feeling that his fireman wanted to say something, but couldn't find the words. Soon curiosity got the best of him and George spoke up. "What's on your mind, Dutch?" There was a short pause. Finally, Blader blurted out. "Buddy, you must be one of those cockeyed optimists. Here you are preparing to assault a 23-foot drift nearly 400 feet long and you don't seem to have the slightest doubt about success." "I'm not even down," replied George, "don't count me out yet. We've got over a hundred men working like Trojans. All I ask is that they give me a trench six feet deep and eight feet wide. If they can complete the job through the second curve, I'll run through that snow like a cat hit in the hind end with a boot jack."

"Hey, Buddy." The voice came from the engineer's side. As George opened his window, the section foreman announced, "We got you that hundred feet." Climbing out, George made a hurried inspection. The clearance was adequate. "Good job," said George as he gave the man a friendly pat on the shoulder.

On his way back to the cab he motioned to Gill with a backup signal. After climbing through the window, George gave three blasts on the whistle.

＊　　＊　　＊

The rules require that when backing the train the conductor control the air brakes using a valve located on the rear of the caboose. Power requirements are signaled to the engineer by means of coded air whistles.

Soon two toots from a high-pitched whistle indicated Conductor Rasmussen was ready. George released the air, moved the Johnson bar in the forward corner and cracked the throttle. After the first exhaust, he heaved her over in the back corner and added another five or six notches on the throttle. Staggering through one complete revolution, both engines threatened to stall. George pulled her wide open and gave a quick shot on the sander. Slowly, almost imperceptibly the train began to gain speed. Once the plow pulled clear of the deep stuff, the train began to move easily. George set the throttle for about 10 mph, but it took an occasional adjustment to compensate for the hills.

George took charge of the train while it passed in front of the depot. Stopping at the coal shed, Blader climbed up on the tank and dumped the chute, while ahead a couple of car lengths, the other fireman did likewise. To fill both tanks with water took less than ten minutes.

George pulled the train ahead until the depot was midway between the engine and the caboose. He set the brake and jumped off the seat box. "Come on, Dutch, let's go get a cup of coffee." Quickly the fireman turned off the blower and anchored the firebox door open. Gill and his fireman met them on the platform, apparently they had the same idea. "Wait up you guys!" yelled Rasmussen, as the conductor and his two brakemen hurried to join them.

Before the men were seated, the waitress had seven mugs of steaming hot coffee lined up on the counter. George tilted his cap back and took a noisy sip. "Right nice of Buddy to treat us. Don't you think so, fellas?" said Rasmussen. All agreed except George. "Swain, you old tightwad. Why don't you surprise us all and pay for the coffee just once?"

Rasmussen finished a long swallow, put his mug down on the counter and leaned over toward George. "All right you ungrateful rascal," replied Rasmussen, "but you'd better get me to Fond du Lac tonight or you have had your last dinner in my caboose."

When they finished the coffee, the conductor paid the bill and the men started back. At the depot, the agent met them in the waiting room. "Say, Buddy, I just got word from Ripon. Half of the townspeople are hiking out to see you make the high dive into the snow." "I'll give them ample warning with the whistle," replied George. To the conductor, he said, "Gill and I will take a little time oiling up the running gear. But when you hear me whistle off,° get on board and be ready for the grand slam."

The two engine crews started for the locomotives. About halfway there, George held out his hand to stop their progress. Looking up into Gill's face, he said. "You have the reputation of being fearless on an engine and I'm not forgetting that you put in many more years at the throttle than I." The comment produced a broad smile on Gill's face. "You realize, that in order to break through, it's going to take everything we can get out of these engines." Nodding his head, Gill indicated agreement. "With your help," continued George, "I'm going to hit that drift just as fast as these engines will run. And I'm not touching the brake valve until we're safe on high ground. Are we together on this?" he added.

Placing a firm hand on George's shoulder Gill replied, "All the way." His voice had the ring of sincerity. "However, I do want to point out that if Ole Granny leaves the rail, she could get turned crosswise. In which case, you and Dutch might wind up with half the train right on top of you."

"I've been watching the way the plow has been riding," returned George, "and I am convinced she'll stay on the track."

°Two short blasts indicating the start of the train.

While the firemen were getting their fires in shape, both engineers ran around the engine with the long spout. Moving up to the front end, George emptied the rest of his oil on the forward journals of the plow. On his way back to the cab he checked his watch. It was 4:10. For a brief moment, he visualized the situation at Ripon. Al Hassman was probably getting No. 9 ready to leave for Wisconsin Rapids. No doubt the work team was preparing to follow him to Fond du Lac.

George began to sense the heavy responsibility of his task. Black smoke was spewing out of both stacks. He looked back to see Gill giving him a high sign, so he returned the wave and quickly mounted to the cab.

The steam gage was nudging the 200-pound mark and the water was about one inch above the bottom of the glass. Looking over to his fireman, he said, "Quick as we get out in the country, we'll give her a good blowing off." Suddenly, the pop valves on the second engine let go. "I guess the boys are ready; you all set?" he asked. Blader gave him a twist of the wrist. George opened the cylinder cocks, moved the Johnson bar in the forward corner and gave two tugs on the whistle cord. Rasmussen gave him a highball. Releasing the air, George took hold of the throttle and eased her out.

After about 300 feet George shut off the cylinder cocks and hooked her back four or five notches. Blader had his injector on and the water level was holding. By the time they reached the open country they were going around 30 mph. George signaled his fireman to open his blowoff cock. Both sides of the engine spewed a 2-inch stream of boiling water along the right of way. After about thirty seconds they shut off. The water had dropped almost out of sight in the gage. Quickly, George put on his injector.

It was nearly six miles to the reverse curve. At around 50 mph,* George eased off to half throttle. Immediately the

---

*These engines were designed to produce their maximum power at around 40 mph. However, as the speed increased the tractive effect began to fall off rapidly. Under ideal conditions, with a light train, a good engineer could get them up to 70 mph.

water began to climb back to the proper level. He shut off his injector and hooked her back another notch. At that speed, it would take less than two miles to have the train rolling full speed. In the meantime he planned to conserve his coal and water.

The track was a little rough and the low joints produced a swaying motion in the cab. Leaning out George noticed the plow was beginning to wallow and pitch. Occasionally it would shift from side to side and shear off a thin slice of snow, sending it back in a fine mist.

Soon the landmarks indicated it was time to start the run. Signaling his intention to Gill with two short blasts, George pulled the throttle wide open. Gill responded with a couple of toots of his own. The roar from his stack indicated he was going all out. Freed from any snow resistance, the engines were soon rocking along close to 60. There was still a mile to go. The speed continued to increase until they were doing around 65.

With the bill of his cap pulled down, George leaned way out. By squinting his eyes, he could see the curve about half a mile ahead. He reached for the whistle cord and gave a series of short blasts. The section men could be seen scrambling for the right of way fence. Blader was busy on the deck heaving coal. George shut the window and shouted, "Hold on Dutch, this is it!" George braced himself against the reverse lever and felt the engine lunge as the pony tracks started into the curve. Instantly, the shock felt like they had run into a stone wall. The impact nearly threw George over the reverse lever against the forward cab window. With snow flying in all directions, the air seemed to explode. George no sooner got back on his seat when the weight of the train came in with a massive thrust, pinning him against the back cushion.

The speed started falling off. At around 45 mph, George moved the Johnson bar forward two notches. The sound of the exhaust coming in from behind indicated Gill was making a similar adjustment.

George saw nearly a hundred spectators lined up along the right of way, some busy taking pictures. An avalanche of snow, flung out from the plow, landed right on top of them.

At 30 mph George added another notch. Using steam at this rate would soon deplete the supply. "Shut off your injector," shouted George. While the engine continued to slow down, George kept dropping the reverse lever a notch at a time. Around 15 mph the reverse lever was in the forward corner. The steam had fallen back 20 pounds.

Suddenly Ole Granny broke through on high ground. As the slack came out of the cars, both engines began to surge forward. Easing off, George brought the train to a stop. Looking back he could see that many of the onlookers were buried up to their armpits. All of them were struggling to free themselves, and the section men quickly ran to their aid. Rasmussen came out and invited them to ride back in the coach. None were hurt. They actually seemed to have enjoyed the experience.

With hat in hand, the conductor gave a big highball. George whistled off and moved the train on into Ripon. From a distance he could see the train order board was out. As he drew closer, it looked like the whole town was crowded on the platform.

Seated at the throttle on No. 9, Al Hassman began to blow his whistle, and the engineer on the work train added his greeting with a series of short toots. As he moved cautiously by the depot, a happy crowd kept George busy returning their friendly waves.

Continuing on down, George waited for a signal to indicate the hind end had cleared the switch which would let No. 9 out. Finally Lauderman stepped off the caboose and signalled to stop. The section men piled off and made a mad dash to catch No. 9.

With drooping shoulders and head down, Blader sat sideways on the seat box. During the run he had exhausted himself trying to keep the engine hot. "Dutch, you did a great job, thanks much," said George. With a weak smile on his face, the fireman raised his head and replied, "I'm sure glad we didn't have another hundred feet of that stuff."

"Did you get a scare when we hit that wall?" asked George. "Well Buddy, let me put it to you this way and you can draw your own conclusion. I'm going to have a dollar and a half laundry bill when we reach Fond du Lac." Both men enjoyed a good hearty laugh.

Moss climbed up into the cab. "Can I ride up here the rest of the way?" he asked. "There are several switches between here and North Fond du Lac that must be thrown and you're just the man to do it," replied George. "What is my conductor doing?" he added. "He's picking up the orders, should be here any minute," answered the brakie.

No. 9 whistled off and started for Wisconsin Rapids. Blader put on his injector and started stoking his fire. By the time he returned to his seat box, Rasmussen climbed up into the cab. "Everybody okay up here?" he inquired. "We're all right, how are things back there?" returned George. "That first jar nearly tipped my stove over. Outside of that we're ship-shape." Reaching in his pocket, the conductor pulled out the orders and gave them to George. After reading them, he handed the flimsies over to the fireman.

What do you say we switch out these hoppers and pick them up on our way back?" asked George. "Good idea," returned Rasmussen. Looking over to Moss he added, "Pull the pin on the hopper back of the second engine. We'll leave 'em both here."

The switching took about fifteen minutes. Rasmussen moved over to George's side and commented, "When we get back on our regular run, I'm going to serve you the finest plate of sauerkraut and pigs' knuckles you ever threw a lip over." With a wave of his hand the conductor started back.

The sun was fast disappearing behind the horizon. Blader had his blower turned up and black smoke was ascending from the stack. "How are we on water?" asked George. "We still have over half a tank," replied Blader. Glancing at the coal pile George concluded they could make North Fond du Lac without any trouble.

"What do you say we yank those boards off, so we can see the rest of the way?" said George. "I'm for it," replied Blader. Both men hopped onto the pilot and stepped up to the running board. By walking along the boiler they were in position to remove the boards with a couple of quick jerks. On the way back to the cab, George caught a high sign from Gill. By the time he took his position at the throttle, Rasmussen had reached the caboose and was giving a highball.

With those heavy hoppers out of the train, they began rolling quickly. The snow rarely reached over two feet. Highballing right on through, George saw the two engines that were stalled on the siding at Rosendale. Four more were parked on a spur at Eldorado.

After uncoupling the train at North Fond du Lac the engines backed on down and picked up Rasmussen and Lauderman. The engine crews headed for the yard office, leaving the locomotives at the coal shed. "Good to see you men," said Washneck as they came through the door.

While George was making out his time slip, the clerk stepped up and handed him a wire. It read as follows:

"To the crews of the Wisconsin Rapids switch runs:
    Please accept my heartfelt gratitude for a difficult job well done.
                    Signed: Jack Rice
                    Div. Superintendent"

"It's not often that Rice gives compliments to anyone," commented Washneck.

Handing the message over to Gill, George commented, "It was those section hands, working like slaves out in that freezing cold who really deserve this message." There was a moment of silence. "You know Buddy, you're absolutely right."

An Album

of Chicago and North Western equipment
in use during Buddy Williams' career

At Van Dyne, Wisconsin No. 2336 is pulling train 297 en route Fond du Lac-Green Bay. This 2-8-2 is the type of locomotive Buddy fired on in years immediately preceding World War I. — *Jim Scribbins*

No. 2908 was a 4-6-2, E-2 shown below pulling *The Mountaineer* out of Milwaukee. Its large tank had been attached when it was converted to oil firing, but when this picture was taken it had been returned to coal, still retaining the large tender. — *Jim Scribbins*

This R-1 No. 376 above is photographed after its arrival at Fond du Lac off the Marshfield passenger. — *Jim Scribbins*

At right Buddy Williams and his fireman, Stan Max, at Wisconsin Rapids in 1952. — *Stanley H. Mailer*

Below, Buddy Williams, now retired, shown with George, Jr. in a picture from the author's collection.

Above, No. 2908 in all its shining beauty was photographed by R. J. Foster and is from the collection of C. T. Felstead. It is the locomotive that "humiliated the Diesel." This print loaned by Harold Stirton.

14.

Competition on the "Hotshot"

Chapter VIII described the circumstances which brought about the decision to put the *Hotshot* into service. Originally, this train was made up of boxcars loaded with paper. Daily, several trains from branch lines delivered their cargo to North Fond du Lac. The paper was quickly switched out on the house track and made ready to depart for Chicago on No. 296, known as the *Hotshot*.

The advantages of speeding freight to its destination proved to be a bonus to the North Western as well as to the paper companies. Swift delivery cut down on inventory and reduced the cost of storage facilities. Soon other industries clamored for the same service. In order to accommodate them, the North Western added merchandise to the *Hotshot's* manifest.

When the Soo Line became aware of the fact that their competitor was getting the lion's share of the business, they decided to do something about it. Since fast delivery seemed to be the key, the Soo Line financed a program to build a fleet of fast freight engines. The performance of these locomotives was to be so much superior to the North Western's Mikados, that winning back the business would be just a matter of time.

The Soo Line was running tests on their first new engine before the North Western became aware of the threat. Fred W. Sargent, President of the Chicago & North Western, called a

177

meeting of the Board of Directors and asked for money to purchase competitive locomotives. These men were a tight fisted lot and when they saw the price tag, they turned thumbs down.

Sargent laid the problem in the laps of his motive power experts. However, the only engines available were the same class of Mikados that they had been using. From an engineering point of view, this was a bit discouraging because the existing schedule was already pushing this engine to the very limit of her performance.

The men with the slide rules got their heads together and came up with a proposal similar to the modification that proved so successful on the famous "400" passenger engines. In view of the fact that the whole project could be handled in the company shops, the cost was not prohibitive and Sargent gave the go-ahead.

In record time the Mikados were completely transformed. New balanced 64-inch drivers replaced the spoke wheels. The boiler pressure was increased ten pounds and the valve ports were enlarged. A mechanical stoker was added, and finally the valve setters came up with their best effort. A coat of black paint gave the engine a new appearance and the testing was started. As was predicted the tractive effort changed very little, but in the speed trials the results were very encouraging. Much to everyone's surprise, this freight engine cranked out 87 miles per hour!

Time was running out and the North Western agents were hard pressed to convince the shippers that they would soon meet or beat the service offered by the Soo Line.

Due to the fact that the *Hotshot* was routed over two different divisions, one engine crew from each division handled the train. A Wisconsin division crew would be heading north on 295, while a Lake Shore division crew was heading south on 296. The following day the situations would be reversed.

The engines, the roadbed, and the tonnage were the same for both crews; therefore, the fastest time for the run would be credited to the best engineer. Needless to say, a strong competitive spirit developed between the two divisions.

For years Len Prunner kept the honors at the Lake Shore division. George shared a little of that limelight while firing for him. Prunner took his pension about the time when the rebuilt Mikados were starting on the stepped-up schedules. George Dix became the engineer and for over a year the Lake Shore division consistently turned in the faster times. His contribution to the success of the run was rewarded by his promotion to traveling engineer. When Dix vacated his position on the *Hotshot,* George bid it in and got it.

Every model of a steam locomotive had its own peculiarities. It required considerable experience before an engineer could take over on a strange engine and get the best performance out of her. Even locomotives of the same class would often require different operating techniques.

At this junction in George's career he had over twenty years at the throttle and he knew just exactly what to expect from each locomotive on his division.

The first two Mikados to be modified were the 2541 and the 2542. Before these engines were reworked, George had run both of them many thousands of miles. However, the day he took over on the *Hotshot,* he found himself on a strange engine. Leaving North Fond du Lac on his first trip, he had 36 loads of paper and 23 boxcars filled with merchandise. Everything was in order when the conductor gave the highball. George eased her out of the yard and when he was sure he had the caboose rolling, he pulled her open. The next four miles would be devoted to getting a good run for the hill at Oakfield. The tonnage rating of each class of locomotive varied with the topography of the route. On hills where the weight of the trains exceeded the capacity of the engine, the engineer would have to double the hill. This meant he would take half the train over at a time. Whenever this occurred, the whole crew was entitled to extra pay. Besides the fact that it usually tied up the railroad, the added cost in wages made such an operation very undesirable from the company's point of view.

Before covering a half mile, George realized that the engine was not pulling as it should. The tonnage of the train was well below the rating for the engine. Being forced to double the hill would prove most embarrassing. Glancing over to the steam gage he noted the pressure was right on the 210-pound marker. George leaned out the window to listen. They were churning along around 20 mph and at this speed the exhaust should have a definite bark. Instead each blast seemed to have a sluggish overlap. The modification of the engine had altered the operating characteristics.

With the old Mikados he knew just where they operated best, but now it was a whole new ball game. Taking hold of the reverse lever, he hooked it back four more notches and the engine began to respond. George had discovered his error just in time to keep from stalling on the hill.

During the remainder of the trip George experimented with various combinations of reverse lever and throttle settings. Before completing the run, he had mastered the technique of getting performance out of the 2542. However, in the process the *Hotshot* fell behind schedule. When he tied up at Chicago, there was a message from the dispatcher requesting an explanation for his delay.

The information given in a delay report can be useful for solving problems which otherwise may continue unchecked. Whenever the reason for the delay would be some failure for which the company was at fault, the report would be exacting and extensive. If there was a remote possibility that the engineer was responsible, the report would tend to be brief and a little vague. Finally after tearing up several starts, George wrote the following delay report:

"LACK OF COOPERATION BETWEEN ENGINEER AND NEW ENGINE YIELDED POOR RESULTS. NEXT TRIP WE'LL BE BETTER ACQUAINTED."

Heading back to North Fond du Lac on No. 295 the next day, George began to realize the potential of this engine. That evening after tying up at North Fond du Lac, George was making out his time slip when the traveling engineer entered the office.

"Hello, Buddy," said Dix. "How did things go on the *Hotshot?*"

George was close to Dix on the Seniority Roster and frequently they followed one another on the same jobs.

"It took me so long to get 'em goin' out of here I thought I'd hafta double the hill," replied George.

"I know exactly what happened," replied Dix. "After my first trip on that engine, I began to wonder if they had ruined a good locomotive."

"You see," continued Dix, "When they rebuilt these Mikados, the valve ports were enlarged and as a result the position of the reverse lever became very sensitive."

"Sounds to me as if we both made the same mistake," said George as he folded his time slip and dropped it in the slot. "So what was your solution?" he added.

Dix struck a match and proceeded to take three or four puffs on a cigar. "You gotta start hooking 'er back much sooner," replied Dix.

"It cost me a delay report but I came up with the same conclusion," returned George.

"Once you get the feel of this engine, you'll like it very much," commented Dix.

As he was speaking, Ben Niland, the roundhouse foreman entered the office. "Well, Buddy, how was your first trip?"

"Ben, all I can say is, I'll hafta do a lot better if I'm going to stay friends with the dispatcher."

Niland walked over to the clerk's desk, picked up a report and started reading. The muffled sound of a distant whistle penetrated the yard office. Dix glanced at his watch and commented, "Looks as if 102 is running late again tonight." After a pause he added, "Guess I had better get on home before the wife starts wondering where I've been."

George went into the adjacent locker room and began to remove his work clothes. The roundhouse foreman followed behind. After planting one foot on a nearby bench, Niland remarked, "I guess you are aware that the general manager gets a daily report on the time 296 arrives in Chicago." George sat down on the bench, crossed one leg over the other and started removing his shoe. Glancing up at Niland he replied,

"I found out about that twenty years ago when I fired the *Hotshot* for Len Prunner. What are you trying to tell me, Ben?"

Niland seemed to be stalling for the want of an appropriate approach to his subject. Finally after shifting his other foot to the bench, Niland began slowly: "The dispatcher kept riding the last two Wisconsin division engineers until they gave up the job. The fact that Dix did so well under the same conditions made the situation all the more embarrassing."

As George continued to prepare for a shower, he interrupted Niland long enough to remark, "Those boys shouldn't feel too bad, after all, there are very few men on either division who can match Dix on a steam engine."

"That's just the problem," returned Niland.

George grabbed his towel and started toward the shower. After adjusting the valves, George stepped under the spray. "What's this problem you're worried about?" shouted George as he proceeded to lather up.

Niland leaned against the doorway and replied, "A Chicago engineer by the name of Turkey O'Brien bid in the *Hotshot*, and I heard by the grapevine that he took the job in order to teach the Lake Shore Wood Choppers how to run an engine."

George pulled the shower curtain aside and stuck his head out. "What did you say that fella's name was?"

Niland raised his voice and answered, "Turkey O'Brien."

After a couple of minutes, George stepped out of the shower and began to towel off. "Ain't that the fellow supposed to be the fastest man on the Wisconsin division?"

"That's him all right and I can tell you he is very fast," answered Niland.

George moved back into the locker room and started dressing. "Well now, I don't see any problem there. If this man is as good as they claim he is, maybe I can learn something from him."

"There's just one thing about this whole setup that bothers me," returned Niland.

"Let's have it," continued George as he pulled on his trousers.

"I'm afraid that you and Turkey will get so involved in beating one another's time that you might be tempted to get reckless."

The comment disturbed George and the tone of his reply showed it. "Ben, you ought to know me better than that."

"I meant no offense," said Niland. "I just don't want to see anyone get into trouble over this thing."

George finished dressing. Rolling his overalls in a bundle he looked over to Niland and commented. "It may come as a disappointment to this fellow Turkey O'Brien, but I run a locomotive to make a living, and I'm not interested in proving who is the fastest man on an engine." With that George snapped the latch closed on his grip and started for the door. "Goodnight, Ben," he added as he departed.

The following day George entered the yard office and as was his custom, he rotated the bulletin board.*

Looking through the glass partition between them, the clerk recognized George studying the board. "Hey Buddy!" said the clerk. "The master mechanic would like to see you in his office."

"Thanks," returned George as he placed his grip on the table.

Hoffman's office was just around the corner. As George passed in back of the clerk, he hesitated long enough to say in a low tone, "Is the ole boy in a good mood?"

The clerk smiled and nodded his head as George continued on to Hoffman's office.

The master mechanic was trimming his fingernails with a pocket knife. Leaning back in the chair he continued the manicure as George stood in the doorway. "Sit down, Buddy," said the official. "This won't take long."

"What can I do for you, Wally?" asked George as he pulled up a chair.

---

*A six-sided configuration mounted on a pivot. By rotating the board, one could read the status of new job assignments, etc.

Hoffman folded his pocket knife, leaned forward and said: "Buddy, I want you to know that I am pleased you bid in the *Hotshot.* Dix did a great job getting that train into Chicago for the early morning delivery, I know I can count on you to do just as well."

George leaned back, crossed one leg over the other and commented: "If you get hold of the dispatcher's report on my first trip, you may not be so confident."

"Don't worry about that," answered Hoffman. "I've already talked to the dispatcher and we both agree you did as well as Dix when he first started." A moment of silence followed. Feeling a little uneasy, George checked his watch and remarked, "I'm on duty in fifteen minutes. Guess I better get moving."

As he started to leave, Hoffman spoke up. "Just one other thing." George hesitated at the doorway. "Don't let this fella Turkey O'Brien goad you into taking any unnecessary chances."

This last comment revealed the real reason for which the master mechanic called him in. At first George resented the implication. After a short pause, George answered in a soft voice. "As you well know, Wally, for over twenty years I've kept a clean record and I intend to have it stay that way."

When George returned to the dressing room, Ed DeBussman, his fireman, was just leaving. "I'll be right along," said George as he hurriedly began changing into his work clothes. Carefully he slid his watch into the bib pocket and anchored the chain through the hole provided for it. A quick swing of his cap and the crown billowed open. With a final tug he adjusted it to the proper angle. Snatching up his grip, he bolted out the door.

The sun had disappeared behind the horizon and the darkening sky was in the process of erasing a residual glow. Walking along the cinder path toward the back of the roundhouse, George reflected on the warnings he had received from his friends. Concluding that they were both unduly alarmed, he attempted to dismiss the subject from his mind.

184

The 2542 was standing on a spur track alongside the south lead to the turntable. As George approached, he noticed that the engine formed a perfect silhouette against the horizon. As he stopped to pull on his gloves, George scanned the view from the pilot to the rear of the tank.

There was a functional symmetry to the lines of this engine that seemed to accentuate brute power. Steam was escaping from around the piston rod and cylinder cocks. An occasional double beat was emitted from the compound air pumps. Tiny rivulets were streaming down the sides of the tank. Apparently, the hostler had just finished taking water and the overflow was still running off.

When he climbed up into the cab, he found the fireman in the process of sweeping the deck. This fellow, DeBussman, couldn't tolerate dirt in any form. As soon as he would get his fire into shape and tend to his lesser duties, he started cleaning the cab interior. With a generous supply of waste (rags) he would start with the face of the gages and wipe off everything in his reach. Windows, arm rests, controls, even the grab irons leading up into the cab were given the once over. When he finished sweeping, he would hose down the deck with hot water from the boiler. DeBussman was just as meticulous about his person. Whenever reporting for duty, he would be clean shaven, wearing freshly laundered overalls and his shoes would be polished to a fine gloss. It was a real pleasure to work with him.

George stepped on the treadle and the firebox doors snapped open. The crown sheet looked in good shape. A thin layer of coal produces more efficient combustion. This condition, however, left very little heat reserve to supply steam in the event of high demand, consequently almost every movement of the throttle called for some adjustment by the fireman. DeBussman had been on the job for over four months and in that time he had become an expert with the stoker.

George lifted the oil can out of the rack and moved it in a circular motion. Satisfied that it was full, he started to get off. Before stepping onto the ground he hollered, "Hey Ed, will you start the dynamo?" DeBussman reached over to the top of the boiler head and opened a steam valve. The whining sound from the turbine-driven generator increased in pitch and the lights came on in the cab.

The malfunction of a part can mean a breakdown. Experience had taught George that time spent preparing the engine was good insurance against failure. While he was poking the long spout in and around the running gear, his practiced eye was searching for any irregularities.

George returned to the cab, took his place on the seat box, and started his routine checks. Air brake, injector, gage cocks, all were in good working order.

The head brakeman climbed up, planted one foot on the bottom board of the coal gate and rested his lantern on his knee. Striking a match on the underside of his pant leg, he raised the glass bowl and lit the wick. After taking his position on the forward end of the fireman's seat box, he greeted George with a wave of his hand.

Reaching overhead, George flipped a couple of switches. Except for the lamps which illuminated the gages, the interior of the cab became bathed in darkness and the headlight cast its beam on the track ahead. "Are you all set?" asked George. DeBussman leaned out and looked both ways. "Let her roll," he returned, as he shut off the blower. George opened the cylinder cocks, released the brake, and eased the throttle out four or five notches. As the engine started to move, a noisy spurt of steam jetted sideways and carried with it the accumulated moisture from the cylinders.

George stopped for several switches and soon they were backing down to couple onto the train. About the time the tender* closed with the train, George spotted the conductor swinging his lantern by his side while making his way toward the engine. The brakeman finished coupling the air hose and George positioned the brake valve so as to charge (up) the train line (air reservoirs).

---

*Coal car, tank and tender are different names for the same thing.

Flipping on the cab light, George stepped over to the locker on the tank, pulled out the water jug and removed the cork. After shaking a cleansing splash out on the deck, he swung it up to the crook of his arm and took a good swig. Before returning it to the locker he repeated the cleansing action and tapped the cork in place.

"Hey Buddy!" George looked down and saw the conductor, Charley Weitzel. The light from the cab appeared to outline his face as he stood peering upward. "Proceed to Fond du Lac and we'll pick up our orders at the depot," said the conductor as he held out his watch. "I've got 9:21 and 30 seconds . . . now!" said Weitzel. "You're only thirty seconds slow," returned George. "Maybe you wouldn't be losing time if you'd get out the key and wind that thing once in awhile," he added. DeBussman stepped over to the gangway with his watch in his hand and broke into the discussion. "Just in case either of you boys are interested in the correct time, I'll be glad to accommodate you."

"Let me know when your fireman learns to tell time and I'll gladly call on him," replied Weitzel. "Better send that troublemaker back to the caboose," returned DeBussman.

During the small talk, the head brakeman was making his way to the switch which headed them out on the main line.

"How much train do we have?" inquired George. "We have sixty loads for a total of 3,100 tons," replied Weitzel.

The brakeman threw the switch and with a long swing of his lantern he signalled to come ahead. "Let's get this show on the road," shouted the conductor as he started toward the caboose.

George made a quick check of the water level, steam pressure, and air gage. Looking over toward his fireman, George said, "Let's go." DeBussman gave him the high sign and started the bell. Reaching overhead George pulled the switch for the cab lights. Except for the high pitched whine of the dynamo, everything seemed strangely quiet in the darkened interior.

A sharp blast of air broke the silence as George moved the brake valve into full release. The groaning of the stubborn

187

brake rigging could be heard echoing down the full length of the train. George moved the reverse lever to the backward motion and gave two short blasts on the whistle. After cracking the throttle (opening it easy) the engine backed in the slack on eight or ten cars. Closing the throttle quickly brought them to a gentle stop. Without hesitating, he flipped on the sander, placed the reverse lever in the forward corner, took hold of the throttle with both hands and pulled it out four or five notches. Before the first couple of exhausts cleared the stack, the engine began to feel the weight of the train. Each time it threatened to stall, George opened the throttle a little further.

Starting a long train must be accomplished with extreme caution. A little carelessness can result in jerking the draw bar clean out of a car and when this happens, there is rarely any doubt as to who is to blame. This engine had an "L" shaped throttle which moved in a vertical plane. The hand hold travelled in an arc which extended fore and aft. It took considerable effort to manipulate this control so as to maintain maximum traction without spinning the drivers (wheels). When starting a long train, the speed must be held to a crawl. If the engineer misjudges the amount of distance required to take the slack, he will be accelerating the engine before the caboose even starts. When that huge mass finally hits the end of the line, the jerk can catapult the occupants from one end of the caboose to the other. Besides being dangerous, such action is not likely to endear the engineer to the boys who ride the rear end.

While the train crossed over onto the main, the head brakeman hopped on and climbed up into the cab. After the engine had traveled seven or eight car lengths, every wheel was turning. George continued to ease the train on down the track. Soon as the caboose rolled on the main line, the rear brakeman realigned the switch and ran to get on.

George faced the rear and leaned out to catch the all aboard signal. Shortly after the flickering light on the caboose came into view, he sighted the long swing of a lantern.

With a couple of yanks on the whistle cord, George looked ahead. The yellow beam bored a hole in the darkness and the two rails appeared to converge in the distance. After hooking the reverse lever back four more notches, he took hold of the throttle with both hands, braced his foot on the boiler head and pulled her wide open.

The 2542 sprang to life. Each thunderous exhaust blasted a plume of smoke high into the evening sky. The powerful thrust from each piston would actually shift the front end of the engine from side to side. Looking along the boiler one received the impression that the iron horse was seeking to take off on a path of its own choosing.

Leaving the south end of the yard, the main line passes through a sea of freight cars. George started whistling for the Scott Street crossing.

The North Western main line is single track both ways out of Fond du Lac. Heading south, the trains take a cross-over which guides them adjacent to the depot platform at Fond du Lac.

The train was rolling about thirty miles per hour as the engine started on the crossover. The sweep of the headlight illuminated the depot.

"The board* is out," shouted DeBussman. Going south the train order board cannot be seen from the engineer's side. George acknowledged the fireman's message with a twist of his wrist.

DeBussman moved over to the gangway and squatted down. While whistling for the crossing at Forest Avenue, George could see someone standing on the platform with train order hoop. When the engine passed by, the agent elevated the hoop and slipped it over the extended arm of the fireman.

Quickly detaching the orders, DeBussman handed them to George who removed the light bulb from the water glass

---

*The train order board is a manually operated semaphore signal. It is used to indicate that the agent has orders from the dispatcher for the train crew.

and held it back of the flimsies* and read them aloud. "All first class trains between Clymon Junction and Fond du Lac have arrived and departed. 296 has rights over 295 to Burnett. Do not exceed twenty-five miles per hour for two miles each side of the siding at Burnett, because of rough track."

George was in the process of returning the light to the cage on the water glass when the fireman shouted, "There's something on the track!" The engine was on the crossover and the long boiler momentarily blocked George's view.

Grabbing the whistle cord, he gave a warning blast. When the engine lined up on the main (line), the crossing came into view. Two hundred feet ahead a 1926 Buick was stalled. The rear wheel was sitting in the trough made for the rail. This vintage car was equipped with large wooden spoked wheels and the driver was trying desperately to move the car by taking hold of the spokes and twisting the wheel.

George slammed the brake valve into emergency and continued with short blasts on the whistle, hoping that the man would get out of the path before they flattened him. He shut off the throttle, dropped the sand and peered anxiously ahead. The back door on the far side of the vehicle opened, a woman jumped out and ran like a deer. The driver seemed determined to save the car even at the cost of his life.

The brakes started to take hold but George knew that they would hit going at least 15 mph.

The man's back would arch and his head drew back as he heaved with all his might. The heavy car appeared to rock slightly and settle back into the groove. The driver took a hasty glance toward the engine. Terror was written on his face and his countenance appeared ghostly in the glare of the headlight. "Oh God help him," muttered George under his breath.

With less than five feet before the impact, the man leaped like a gazelle. The pilot crunched against the rear end of the Buick and sent it spinning.

As the engine continued on past the crossing, George could see the man's form writhing alongside the track. Before

---

*The orders are written on this transparent paper so that light passing through outlines the message.

the train came to a stop, he whistled out a flagman,* climbed off and raced back to the crossing.

The man was lying face down. Carefully George rolled him over. His eyes were wide open in a fixed stare. Running his hands over his body and down along his legs, George was relieved to find he was all in one piece.

Apparently the man was in shock. After a few seconds he spoke in a trembling voice, "I thought I was a goner." "Can you move your legs?" asked George. The man responded by trying to get up.

DeBussman took hold under one arm and George took the other as the man got to his feet.

The wigwag was swinging and its bell was dinging. Headlights and beeping horns indicated several cars were waiting at the crossing. George started to help the man toward the depot, but about that time the pop valves let go, spouting a column of steam skyward. "Take care of her," said George, "I can handle this." DeBussman hurried back to the engine.

The man no sooner sat down in the waiting room when in came Conductor Weitzel all out of breath. "What happened, Buddy?" "We hit this fellow's car and shook him up a bit."

Weitzel whipped out a pad and pencil and started taking down the necessary data for the accident report. "Did anyone else see this thing?" he asked. "There was a woman who got out of the back seat and ran just before we hit," replied George. "Let me have her name and address," returned Weitzel. The man's eyes flashed and his jaw seemed to set. "That's none of your d - - - business."** "Come on now, let's have it," said Weitzel. The man just sat there looking daggers at the conductor. George interrupted, "Tell me, Fella, why in the world did you wait so long before you got into the clear?" "My finger got caught in the brake drum and I couldn't pull it loose," he answered apologetically. With that he held out his left hand. The end of his index finger was a bloody mess.

---

*One long and three short blasts.

**Investigation revealed a case of hanky panky between the woman and the driver — hence his reluctance to identify her.

191

Weitzel put on his glasses, took one look, and went over to the ticket agent. "Call an ambulance for this fellow and send him to the hospital."

George glanced at his watch, "What do you want to do, Charley?" Weitzel rubbed his chin and said, "Let's call the dispatcher and see if he'll let us proceed with the same orders."

After a short explanation the dispatcher said, "Can you start from there and make Oakfield Hill without doubling?" "Hold on," returned Weitzel, "I'll ask my engineer." George considered the question for a moment and answered, "Tell him I'll put the train over the hill if I hafta get off and shove it myself." The dispatcher was satisfied and gave the go ahead.

"Let's get with it," shouted Weitzel as he started back to the caboose.

On the way out of the depot, George turned back and said to the injured man, "Good luck."

As he made his way to the cab he could hear the horns of impatient motorists still waiting at the crossing.

Back on the engine, George noticed that DeBussman had wisely moved the brake valve into the service position.

George climbed up on the seat box, grabbed the whistle cord and called in the rear flagman.* The fireman adjusted the stoker and inquired, "How is he?" Moving the brake valve into full release, George waited till after the blowdown to reply. "He's got a nasty looking finger, outside of that he may have a few bruises."

DeBussman anchored the firebox door half open as the pops began to sizzle. "Well, what do you say, Buddy? Are we going to back down and take another run for this hill?" George placed the reverse lever in the back up motion. "The dispatcher agreed to let us run on existing orders if I promised to make Oakfield without doubling the hill." As George reached for the throttle, DeBussman commented, "That's a big order, but I believe you're just the one that can do it."

George backed the engine until the slack was in. Then without shutting off, he put the reverse lever in the forward

---

*Five long blasts of the whistle is signal for rear flagman to return to cab.

corner and started jockeying out the throttle. Once the entire train was moving, he eased off just enough to keep it rolling. When he caught the all aboard signal, George pulled the throttle wide open. By hooking her up a notch at a time, the exhaust started cracking like a well tuned passenger engine. "That's it," said George approvingly, "Keep talking to me baby and we'll put this train over the hill." DeBussman knew what was expected of him and he had the steam right up under the pops.

The train was traveling close to forty miles per hour as the grade began to steepen. In short time, the heavy tonnage made itself felt. Grudgingly George dropped the reverse lever a notch at a time.

Nearing the top, the train was barely moving. The engine threatened to stall after each exhaust, but somehow, the 2542 continued to grind away at this pace until they were over the hump.

Noting the definite increase in speed, DeBussman shouted, "Buddy, you had me worried for awhile." George hooked her back a couple of notches and hollered back, "I'm going to like this engine."

DeBussman got down on the deck and checked on the condition of his fire. Moving over to George's side he said, "Turkey O'Brien will be chafing at the bit by the time we get to Burnett."

George pulled out his watch and studied it for a moment. "Looks like we're going to stick* him for at least thirty minutes," he replied.

The crescendo from the stack was fast becoming a roar and soon they were clipping along sixty-five and seventy mph.

The North Western cuts through the outskirts of Waupun, the home of the state penitentiary. The sky was clear and the stars were out in profusion. Looking back through the haze of the trailing dust, George could see the lights of the prison compound.

---

*No. 296 couldn't move out on the main line until No. 295 went by. Sticking No. 295 meant forcing them to wait beyond the scheduled time for the meet.

To charge through the countryside on a good engine with sixty loads on behind is a thrill that never grows old. With his eyes trained on the track ahead and his hand resting on the throttle, George delighted in pitting his skill against the contour of the land. Every hill and curve presented a challenge to his judgment.

As each engineer knows, that huge boiler out in front is under terrific pressure. While heading up and down the sags, the water surges fore and aft. If for any reason the fireman fails to keep the boiler filled to a safe level, the water may shift enough to leave the crown sheet bare. In just a few seconds the dry area becomes white hot. When the water flows back on that surface, it turns into steam instantaneously. Most boilers will take twice the working pressure before blowing up. But when they do let go, the result is devastating and the engine crew is almost always killed.

Clymon Junction is the first stop 296 makes for water. The penstock is located at the south end of the depot. A quarter mile beyond that, the St. Paul Railroad intersects the North Western. A crossing tender operates the signal which governs the passage of all trains.

A mile out of Clymon Junction, George eased off (on the throttle) and held the whistle open for one long blast. After adjusting the brake for a fifteen pound application, he lapped* the brake and released the independent (engine) brake.

Up till now, this writing has emphasized the importance of power. However, in order to clear up what may be a false impression, consider the following. The skill required to stop a train safely is far more important than the ability to maximize the power of a locomotive. Every time an engineer takes hold of a brake valve, he must consider the effect of many variables. Speed, distance, topography, weather, are a few.

Some hogheads were fearless, others were cautious. It was that rare combination of skill and judgment which allowed a few engineers to be both.

---

*Positioned the control to hold the pressure in the brake cylinders constant.

As they approached the depot, the train order board swung out. George acknowledged the signal with the whistle and DeBussman prepared to pick up the orders.

After easing the train to the south end of the platform, George managed to spot the tank under the spout.

DeBussman handed over the orders and George removed his gloves and read them aloud: "295 take siding at Burnett for 296. All first class trains between Clymon Junction and Butler have arrived and departed." While the fireman was taking water, George made the rounds with the long spout. Upon returning to the cab, he stowed the oil can and wiped off his gloves.

DeBussman climbed aboard and announced, "The tank is full and I'm ready for ya."

Looking ahead, George could see the gate at the crossing was against him. With a couple of tugs on the whistle, George looked over toward his fireman and suggested, "Let's go on down and wait at the crossing. DeBussman nodded his head and George moved the train to within a car length of the intersection.

Standing in the glare of the headlight with his red lantern in hand, George recognized Jake the gate tender. He was a little man with a ready wit and a keen eye, who had worked in this location for nearly forty years and knew the engine crews of both railroads by their first names. His was a lonely job and Jake always welcomed the opportunity to strike up a conversation.

Walking over to the fireman's side of the cab, Jake shouted, "The St. Paul is running a passenger extra. They should be along any minute now."

George put on the lights in the cab and moved over to the gangway. He looked down at Jake and said jokingly, "You should have stopped the St. Paul train and let us by." "That you, Buddy?" asked Jake. "Come on up and see for yourself," replied George.

Jake set his lantern down on the cinders and climbed up into the cab. Breathing heavily from the exertion, he commented, "Kinda thought I recognized your whistle. By the way, who is the new flash running opposite you?"

DeBussman finished taking a drink from the jug and handed it to George. "He's a Chicago man by the name of Turkey O'Brien." After taking a long swig, he returned the jug to the locker. "What made you ask?" added George. "Going north the other night, he came barreling down so fast that I was afraid he wouldn't get stopped for the crossing. When I opened the gate to let him by, he cussed me out for not moving fast enough," answered Jake. DeBussman laughed and commented, "Must be a real nice guy."

George planted a foot on the bottom board to the coal gate. "I'll tell ya what's the matter with the guy," said George. "He's made his brags about teaching us Wisconsin wood choppers how to run an engine and it looks like he's getting a little desperate."

The headlight of the St. Paul engine showed up in the distance. "Gotta go," said Jake as he hastened to get off.

After the passenger rumbled by, Jake opened the gate and gave the all clear. George put out the cab light and looked over to his fireman. DeBussman gave him the high sign and George started the train.

Glancing at his watch, George noted he was running thirty minutes behind schedule. Hopefully, with no more delays, he could yet make up ten or fifteen minutes.

Continuing on south, the hills started to flatten out. The 2542 began stepping along between 60 and 70 mph. On the long curves to the right, George looked back to inspect the train. In this way a hot box can often be detected. An occasional spark could be seen as the flanges ground along the rail.

Small farming communities are located all along the route. George was careful to give them advance warning with the whistle.

Scanning the horizon, George detected several landmarks which informed him they were nearing Burnett. He eased off on the throttle and let the train drift. Recalling the twenty-five mph slow order, George started his first brake application.

While rounding a long curve, an amber signal* could be seen in the distance. "45" shouted the fireman. "45" repeated George, as he eased off a little more. Soon a light could be seen. Apparently 295 was just heading into the siding at Burnett.

The following signal was red. "Redeye,"† shouted DeBussman. "Redeye," returned George as he added another ten pounds in the brake cylinders.

DeBussman moved over to George's side. "What have we got here, a cornfield meet?"‡ he inquired. George leaned against the back rest and readjusted his cap. "That's the way she looks to me," he commented.

George was preparing to make a full stop when the red signal turned green. The headlight facing them blinked off and came back on dim.** George released the brake, pulled out the throttle and advanced the reverse lever six or seven notches. The 2542 started to bark with renewed vigor.

As the two engines passed, George got his first glimpse of Turkey O'Brien. The man had a huge handlebar mustache and he wore his long bill cap backwards.

Standing in the gangway on the engineer's side, DeBussman took in the same view. "Turkey is a regular Barney Oldfield," remarked George. DeBussman roared with laughter. "Why say, that fella looks like he's going 90 mph just sitting there," added DeBussman.

From there on during the remainder of the trip everything seemed to click and George concentrated on making up the delay.

As they pulled into the freight yard at Chicago, DeBussman looked at his watch and announced, "Buddy, did you know we picked up twenty minutes between here and Bur-

---

*Amber light by night is the same as a 45° position of the arm on the semaphore and signifies that the following block may be occupied.

†Red light by night is the same as 90° position of the arm on the semaphore and means stop.

‡Two trains on the same track heading for each other.

**Blinking the headlight at an oncoming train indicates the train is in the clear on the siding and the switch is aligned for the main line.

nett?" George gave his fireman a broad smile and answered, "If this engine keeps on talking to me as she did tonight, Turkey O'Brien is going to have his work cut out for him."

George was rapidly developing a genuine affection for the 2542 and he made up his mind to see to it that she received the proper care. Much like an expert auto mechanic, George had trained himself to detect trouble by listening to his engine run. Every strange click, pound or hiss became a clue. At the end of each trip George would make out the engine report and in it he would be careful to describe every detail that required attention. Since it was the only motive power capable of handling the *Hotshot,* the officials were most cooperative. Consequently, the 2542 was always operating at its performance peak.

After a couple of weeks on the *Hotshot* George located landmarks along the route which prompted him to perform some operation. A huge elm tree halfway up Oakfield Hill became the ideal spot to move the reverse lever forward two notches. The tall steeple of a country church marked the location for the start of a brake application. Nearly every trip he managed to clip off a minute or two from the previous best time.

Meanwhile, Turkey was having problems. Apparently he never quite understood the need for hooking her up early on the start.* As a result, his fireman found it almost impossible to keep him in steam and Turkey was consistently running behind schedule. The situation was becoming embarrassing and he started to blame his crew for the delays.

One evening while making out his time slip at North Fond du Lac, the clerk couldn't resist chiding him for his boasting. "We heard you took this job to teach the wood choppers how to run an engine." Turkey pretended not to hear him. The clerk continued, "I've been checking on the reports and I noticed Williams has been beating your time by nearly a half hour every trip."

---

*"Hooking her up" refers to moving the reverse lever toward top center on the quadrant.

Turkey slammed his pencil down and replied, "I'll beat his time if I have to put the whole train in a ditch." With that he picked up his grip and stormed out of the office.

From that time on, he began to ignore slow orders and to show complete disregard for safe operating practices. One night he was traveling over sixty on a twenty-five mile per hour slow order track when a brakeman, standing on the rear platform of the caboose, was nearly thrown off. At this point, the conductor realized that the man would kill someone if he remained at the throttle, so he reported the incident to Jack Rice, the superintendent.

Turkey promptly received a wire to appear at the superintendent's office. After questioning him at length, Rice concluded that O'Brien was a poor risk and told him to go back to Chicago because he would never again run an engine on his division.

When George found out that Turkey was barred from the Lake Shore division, he recalled to mind the concern expressed by Ben Niland and Walter Hoffman. Maybe if they had talked to Turkey as they did to me, reasoned George, he would not have lost his head.

The fact that Turkey O'Brien was no longer competing for the honors made not the slightest difference in George's performance on the *Hotshot,* "Because," as he told Ben Niland, "he ran an engine to make a living and didn't need to prove who was the fastest man."

15.

Humiliating the Diesel

There was an interim period between 1930 and 1940 when the outcome of the struggle between steam and Diesel hung in the balance. It was during this time that the Chicago and North Western introduced the famous *400*. Billed as the train that set the pace for the world, the *400* was a high-speed run between Chicago and the twin cities, Minneapolis-St. Paul. The schedule called for 410 miles in 390 minutes. Twelve of their Pacific type locomotives were modified to pull this fast train. The original 75-inch drivers were replaced with 79-inch specially balanced wheels. The boilers were beefed up to handle a 25-pound increase in the working steam pressure. The valve gear was reworked and adjusted for high-speed operation.

In a test conducted by the Interstate Commerce Commission, one of these locomotives pulled a four car train eight miles in four minutes flat. Among these engines, one bearing the number 2908 proved to be exceptionally fast and it soon became a favorite with the engineers. On one occasion, when the *400* was running late, the 2908 made the trip from Milwaukee to Chicago, a distance of 86 miles in 65 minutes.

Although these engines proved to be fast and reliable, the Diesel came out ahead on economy. One by one, as the North Western took delivery of the streamliners, the steamers were put on stand-by. By 1948 the transition to Diesel was nearly complete and the sound of a steam whistle was fast becoming a memory.

Occasionally, a Diesel would break down and one of the old reliables would be fired up to save the day. Whenever this occurred, the word got around quickly and everyone turned out to see the steam engine doublehead the streamliner.

It was early on a Saturday afternoon in the fall of the year. George had just completed a round trip on the De Pere and Little Rapids switch run. The work load that day was lighter than usual and the whole crew was looking forward to hearing the last half of the Green Bay Packers versus Chicago Bears football game. After spotting the train at the south end of the Green Bay yard, George and his fireman took their Diesel on to the roundhouse. Upon completing the procedure for stowing the Diesel, George grabbed his grip and started for the yard office. As he passed the turntable, he noticed the No. 2908 was standing in front of the water tank and several workmen were busy getting her ready for something. George stopped just to admire the old race horse.

In a moment of reflection, he recalled some of the opportunities he had had to put that high wheeler through her paces. This particular engine seemed to have a personality all of her own. Being stoker-fed and a good steamer, George found that the harder he worked her, the better she responded, and the faster she went the stronger she became. George had her on No. 209 for nearly three months and almost every night he would be making up time between Milwaukee and Green Bay.

With a parting glance at the rear end of the tank, George noticed the paint was pealing badly. The large numbers "2908" were barely readable. The couplers were covered with a coat of rust, mute testimony to their inactivity. There was a time when that engine never left the roundhouse unless she glistened like a black diamond. In the days when the *400* was running competition with the St. Paul's *Hiawatha* and the Burlington's *Zephyr*, the North Western dressed this engine up with a large blue electrically lighted marquee. On the front end, this decorative emblem spelled out the *400* in ten-inch white letters. Just ahead of the cab, under the running boards on each side, a large sign repeated the message. On the cover of each cylinder was a six-inch shiny gold star, which signified that the regular engineer had over 25 years of experience. The unique mars headlight graced the top of the smokebox and pointed her warning beam skyward. Admirers would be lined up along the track as she sped on by.

To the younger engineers, the 2908 might be just another hunk of iron awaiting the scrap heap, but George preferred to remember her as the locomotive that set the pace for the world.

Continuing on to the washroom, he removed his overalls and proceeded to clean up. As he rinsed the soap off his face, the callboy entered, seeming a bit excited and out of breath. "Am I ever glad to see you!" he said.

"What's the problem?" asked George.

"One of the units on the 216 Diesel went dead and the other is overheating badly. The dispatcher figures they will be about 15 minutes late getting in here," answered the callboy.

George sensed that he was about to relinquish his nice quiet evening. "Why don't you get someone off the extra board?" he asked hopefully.

"We tried, but no one could make it on such short notice. Mr. Hoffman sent me to ask you to doublehead 216 as a personal favor to him."

Walter Hoffman was the master mechanic and it was his responsibility to provide the motive power that kept the trains rolling. No. 216 was on a mile-a-minute schedule, with important connections in Chicago for passengers heading east. When George hired out firing, it was Walter Hoffman who assigned him to fire for his father, and George never forgot the kindness.

"How much time do I have?" asked George.

"No. 216 will arrive in about 20 minutes," replied the callboy.

"Have you got a fireman for me?"

"Fireman Pagel is out back of the roundhouse getting the 2908 ready."

George put his overalls back on. "You can call Wally Hoffman and tell him that Buddy Williams said for him to go back and finish listening to his football game. The 206 will arrive in Milwaukee on time." With that, George grabbed his grip and took off. When he arrived at the engine, fireman Pagel was on the back of the tank taking water.

George climbed up into the cab, grabbed the oil can and started to oil the running gear. While squirting oil on the guides, a workman came alongside and said, "I think she is ready for you, Buddy."

George thanked him, but continued around the engine just to be sure nothing was missed. When he returned to the cab, he opened the firebox door and inspected the crown sheet. Being an old engine, any weakness in this area could be disastrous. George intended to lay everything this engine had right on the rails.

The heavy cover on the water tank banged closed. The fireman came over the coal pile, grabbed the curtain rod and swung down onto the deck. George tested his injector* and checked his gage cocks.** Pagel moved over to the engineer's side and said, "I'm all set, Buddy."

Looking down at his fireman, George said, "I sent word to the master mechanic that 216 will arrive in Milwaukee on time. Now I intend to keep that promise even if I have to tear the whole front end off this engine in the process."

George studied Pagel's reaction and added, "I'm going to need all the help I can get. Are you with me?"

George had a reputation as a very fast man with an engine, and there were a few of the younger firemen who had expressed some fear of working with him.

A big grin came over Pagel's face. Patting George on the knee, he said, "Buddy, you crack the whip and I'll make the trip." The answer seemed especially appropriate and George felt assured that he could count on him.

---

*The engineer's injector provided water for the boiler when the demand was more than the fireman's injector could handle, and was used in case of any failure of the fireman's injector.

**Three valves, each one was located on a level about three inches lower than the other. By opening the top valve, steam or water would come out of the opening, depending upon the level of water in the boiler. If the lowest valve showed steam instead of water, the engineer would know he was dangerously low on water. This system is a double check on the water glass indicator.

"Start the bell," shouted George, as he leaned out the window. Much to his surprise, the whole roundhouse crew came out to give him a send-off.

The coming of the Diesel had meant the eventual loss of jobs to many of these men, and they thoroughly enjoyed seeing the streamliner humiliated. One of the boilermakers hollered out, "Hey, Buddy, we are depending on you to make them guys ashamed that they ever bought those coffee grinders."

George opened the throttle and waved back as the engine moved out briskly. Passengers awaiting the arrival of the streamliner were milling about the platform. The sight of a steam locomotive chugging on by attracted their attention.

As the engine came to a stop at the south end of the depot, George spotted Bob Dilly standing on the platform with grip in hand. The streamliners changed crews here and Engineer Dilly was the regular man from Green Bay to Milwaukee. As an engineman, Dilly was known to be a steady reliable runner, but he was nearing his pension and as was the case with many of the older men, he preferred to let the pace-setting be done by the younger engineers.

George climbed down and went over to greet Dilly. "Howdy, Bob."

"Hello, Buddy," returned Dilly and they shook hands.

"What seems to be the trouble with the Diesel?" inquired George.

"The electric-drive motors are overheating and at a certain temperature, the circuit breakers cut the power off completely," replied Dilly. "If I understand the dispatcher's report correctly, we will have to depend on the steam engine to get us to Milwaukee," he added.

When double-heading, the engineer on the lead engine is in charge of handling the train. Normally, the regular man accepts this responsibility, but Dilly had no intention of giving up his comfortable swivel chair for a seat box on a rough-riding steam engine. Turning to George, he said, "You know, my back can't take the jarring like it used to. Would you mind taking the head engine?"

George assured Dilly that he would be glad to accommodate him.

The air horn on the Diesel interrupted the conversation and alerted everyone that the streamliner was coming in. George and Dilly both pulled out their watches. "We are going to be 15 minutes late out of here," commented Dilly. "I don't see any need to be tearing up the rails. After all, they can't expect too much out of that old steam engine. Now, when you approach the curves," continued Dilly, "just look back and I'll put my hand out to indicate where you should start holding 'em up." *

An engineer who has held a particular job for an extended period, develops an almost fatherly sense of responsibility for everything connected with his run. Being well aware of George's reputation, Dilly felt a little uneasy at the prospect of being a spectator while his train was being rocked around the curves at high speed.

With the clanging of a nervous bell, the Diesel eased to a stop about a hundred feet behind the 2908. Before starting back to the cab, George shouted, "Don't worry about that brake valve, Bob, just get up there in your rocking chair and relax." George knew every bend in the rail and he had no intention of riding backwards in order to receive instructions on how to operate the brakes.

George no sooner sat down on the seat box when the head brakeman signaled to back in and couple up. The 2908 was equipped with a powered reverse gear. This control replaced the huge clumsy Johnson bar. It was a significant improvement from the standpoint of safety and it could be operated with one hand.

After releasing the engine brake, George moved the reverse lever into the back-up position, cracked the throttle and eased the engine back until the tank made contact with the Diesel. When the brakeman finished cutting in the trainline,** George checked the results on the air gage. Then he saw Bill

---

* "Holding 'em up" refers to braking action.
** Connecting up the air brake hose between the two locomotives.

Wood, the conductor, threading his way through the passengers, with the orders in his hand. George climbed off to meet the conductor.

"Well, if it ain't ole Spark Plug himself. They must be scraping the bottom of the barrel these days," said Wood with a chuckle.

"It's about time they do a little scraping," returned George. "The master mechanic told me you boys forgot how to read a time card." *

George had been on a freight regularly for over a year and had lost contact with the passenger train crews. But if he could have hand-picked his crew that day, he wouldn't have changed a one of them. Conductor Wood was a rough and ready outgoing personality, with a good sense of humor. When it came to railroading, he possessed a serious dedication to his work.

"I guess you are aware of the trouble we have been having with the Diesels," said Wood.

"I understand that every time 216 is late, the Chicago officials send Hoffman a nasty wire. Let's see if we can spare him another bawling out today," George returned.

"We have rights over everything all the way to Milwaukee and the way freight will be in the hole** for us at Appleton Junction," said Wood, as he handed over the orders.

After comparing watches, George observed that the loading was about finished. Wood started to leave, saying, "Grab 'er right by the ears, Buddy; don't let up on her and I'll give you all the help I can."

While climbing up into the cab, George hesitated, glanced back and shouted, "Hey, Bill!"

The conductor had progressed to the far end of the tank. Stopping in his tracks, Wood turned and looked back.

"Be sure you have a good hold on the grab-iron before you give the highball," shouted George, "or I'll leave you standing at the depot."

---

* An official railroad publication which gives the schedule of all trains on a division.

** On the siding in the clear.

The conductor smiled, gave him an understanding wave and hurried on back. Looking down on the platform, George could see Walter Hoffman hurrying toward the engine. "Thought I sent you back to your football game," said George.

"I got your message, Buddy, and thanks much for helping me out."

About that time the conductor gave the highball. George released the air and opened the throttle. Hoffman hollered, "Don't forget these old engines are not supposed to be run over 85 mph!"

The train started without taking the slack. After George adjusted the throttle, he looked down at the master mechanic and pretended he didn't hear him. Hoffman attempted to run alongside and repeat the message. George cupped his hand back of his ear and feigned a puzzled expression. Hoffman was about to try again when the drivers started to slip and the noisy exhaust drowned him out. Finally, he gave up and just stood there with his hands on his hips and an exasperated look on his face.

As the engine passed over the crossing on the south end of the depot, Pagel shouted, "What was Hoffman trying to tell you?"

George nudged the throttle out a little farther and replied, "Wally's worried about me running this engine over 85 mph."

Pagel beamed a big grin and added, "Wonder where he ever got an idea like that?"

A column of black smoke jetted skyward as the massive drive wheels clawed at the rails. After adjusting the reverse lever back three or four notches, the exhaust sharpened considerably. George watched the air gage as he made a light test application. Soon as he sensed the braking action, he released the air brake, waited for the blow-down and reset the valve in the running position.

Going south out of Green Bay, the North Western tracks pass over numerous intersections. The unfamiliar sound of the steam whistle was attracting attention all along the right of way. The air horn on a Diesel operates with a small tab.

Neither the tone nor the sound level can be controlled, whereas with the steam whistle one could almost play a tune. Many engineers developed their own trademark with the whistle cord. Except for a couple of extra trills on the end of the last toot, George stayed with regulations.

Half a mile from the Green Bay depot a signal protects the St. Paul's crossing. The semaphore was straight up. The speed limit for the crossing was 40 mph.* The train was doing better than 50 as they rambled over the St. Paul track. The blast from the stack was becoming a constant roar. The 2908 was starting to show her mettle. By the time they passed the little station at De Pere, the train was stepping along close to 70. As the engine headed into a long gentle turn, George looked back and watched the hind end slide gracefully around the remainder of the curve. Satisfied that everything appeared normal, he returned to the whistle cord and blew the warning signal for each crossing.

As long as the fireman knew his business with a stoker, an engineer could work the engine harder without knocking back his steam. Pagel was doing a good job and the pressure never varied over five pounds from the time they left the depot.

As the engine continued to gain speed, the little farming community of Wrightstown slipped by. George leaned out into the wind to make a visual check on the running gear. The side rods were just a blur. In spite of the powerful blast from the exhaust, the wind flattened the smoke less than six feet from the stack and a thin ribbon of black smoke trailed over the train.

There was a reverse curve on the approach to a little siding called Sandcut. About 400 feet before reaching the tangent, George grabbed off ten pounds of air and held the independent brake** in full release. This method of applying the brakes has the effect of stretching the coaches out tight

---

*It was not uncommon for an engineer to exceed speed limits whenever such limits were known to be conservative.

**Engine brake only.

while the engine continues to pull. The result is, the coaches lean into the curve, but rocking is kept to a minimum, and very little speed is sacrificed. The big boiler lunged as the lead trucks took up the side thrust and guided the engine around the curve. George chuckled to himself as he pictured Dilly's arm waving from the cab of the streamliner.

The depot at Kaukauna was situated on a curve at the west side of the city. The train order board* went clear just as the station came into view. Two short toots informed the operator that the message was received. The agent came out on the platform to give a friendly wave. But when he realized how fast the train was approaching, he made a mad dash back into the depot and slammed the door. The little episode gave both Pagel and George a hearty laugh.

About a mile out of Appleton George eased off on the throttle, made a 12-pound reduction and lapped the brake valve. Soon he felt the train slowing down. Before it came to a complete stop, the automatic steps folded out and the conductor stepped off. Everyone moved with a sense of urgency. In less than a minute George got the highball and they were on their way out of town. Looking over to the fireman, George shouted, "How we doing?"

Pagel glanced at his watch and hollered, "Keep it up, Buddy, we're gaining on it!"

Appleton Junction was three miles ahead. The way freight would be waiting on the siding. It was uphill and the engine labored to gain speed. The train order board was clear at the junction and the station platform became visible on the engineer's side. George called Pagel over to see the welcoming committee. The way freight crew and the station agent lined up on the platform and gave out with a lusty highball as they thundered by. George returned the wave and thanked them with a couple of tugs on the whistle cord. Before returning to his side, the fireman commented, "Those boys must have been following the dispatcher's report on our progress."

---

*A semaphore manually operated by the station agent to indicate whether or not there were orders to be picked up by the oncoming train.

"Looks like they're trying to make a couple of celebrities out of us," returned George with a smile.

Neenah, a wealthy little town of about 10,000 people was seven miles away. Heading down a sag, George hooked her back a couple more notches. The short distance hardly gave enough room to really get the train rolling.

Right after passing over the long bridge at the entrance to town, the North Western tracks cross over both the Soo Line and the St. Paul. The speed limit is 30 mph. George began braking just before the engine started on the bridge. The signal for the crossing was clear. By the time they were pounding over the intersection the speed was down to 40 mph.

Loading the train was completed with dispatch and in about ninety seconds the conductor swung his arm in a long arc and they were soon highballing out of town.

Next stop, Oshkosh, was 18 miles away. State Hospital, a flagstop, was midway between Neenah and Oshkosh. At that point, the train was streaking along at better than 90 mph. The 2908 seemed eager to run and George just let her have the reins.

Due to the fact that the North Western right of way is all fenced off coming into Oshkosh, the danger to personnel is minimized. Two miles out of town, George eased off on the throttle, pulled the whistle wide open, held it there while he grabbed off 15 pounds of air and lapped the brake valve. The train was still going better than 60 mph as they entered the city limits. About a quarter of a mile from the depot, George added another 10 pounds in the brake cylinders. Just before coming to a complete stop, George released the independent brake (engine brake) and reduced the trainline down to ten pounds. With the engine still pulling lightly, the actual stop was almost imperceptible.

As he moved the brake valve to charge the trainline, George pulled out his watch. Looking over to his fireman, he said, "We got two more stops, 86 miles to go and 76 minutes to make it."

Pagel's eyes twinkled. He was obviously enjoying himself immensely. After reaching over to open the blower valve, he commented, "If this ole gal holds together till we get to Mil-waukee, we'll have a record for the Diesels to shoot at."

"Let's hope Dilly doesn't have a heart attack before we get there," returned George with a big grin.

The conductor's highball cut the conversation short. It was slightly uphill for the next six miles. Half a mile out of town, the signal for the swing bridge came into view on the fireman's side. "Clear!" shouted Pagel. George pulled her wide open.

The train rumbled over the bridge and headed for Fond du Lac, 16 miles away. At intervals, George leaned out the window to listen. Each adjustment on the reverse lever was carefully tuned to the message he received from the exhaust. The train was doing around 70 mph as they passed Van Dyne, a little station near the top of the grade. Tower DX controls the entrance to the huge freight yard at North Fond du Lac. It is downhill from there all the way to Fond du Lac. The speed increased until the engine began to vibrate all over. Every loose joint or part in the cab began to rattle and bang until it almost drowned out the exhaust.

Looking ahead, George could see several figures on the footpath to the roundhouse. Grabbing the whistle cord, he gave them a lengthy warning. As they came closer, George recognized a good many of the roundhouse crew. Each was giving an encouraging wave. George thanked them with a couple of toots as they roared on by. A mile and a half from the station, George held the engine brake in full release and grabbed off 12 pounds of air on the trainline brake. While waiting for the brakes to take hold he eased the throttle to a drifting position and started whistling for the Scott Street crossing. Beyond the crossing there was a curve to the left, followed by a bridge over the Fond du Lac river.

About two blocks from the depot, Pagel started the bell and signalled George that he was on his way over the top of the tank in order to get ready to take water. George added another ten pounds of air and lapped the brake valve. By using the throttle to drag the train to the precise location, he came to a stop with the opening in the tank directly across from the penstock.

Picking up the oil can, George took hold of the handrail and slid all the way to the ground, barely touching the steps on the way. After aligning the water spout on the penstock over the tank, George proceeded to oil every moving part that he could reach, the trailing trucks, driving boxes, lateral motion plates, piston rod and guides. As he was finishing up on the left side, conductor Wood hurried up to the engine with his watch in his hand to say, "You must have fed this ole horse a big bucket of oats before leaving Green Bay."

"I've been beating her over the back pretty hard," said George. "How are you coming with the loading?"

"Soon as you're through with your oiling, we'll be ready for you," replied Wood.

George gave him a nod and started around the front of the engine. When he returned to the cab, Pagel had his blower on and the stoker working. Black smoke rose from the stack. Suddenly, the pop valves let go and a thin pencil of steam spouted skyward. Quickly, Pagel cut in his injector. George positioned the reverse lever in the forward motion and stepped over to the gangway on the fireman's side to look back. The conductor gave the highball as the wagons started moving away. Before he reached his seat, George flipped the brake valve in full release and pulled the throttle half open. From here on, the cards were down. The next eight miles were all upgrade; with eight cars and two dead Diesel units he would do well just to hold his own.

George had been working the 2908 unmercifully, but now it was going to get a murderous beating. Soon the throttle was wide open and he began to hook her back slowly. The blast of the exhaust sounded as if it would tear the stack off. For awhile, the water level was dropping even with both injectors on. George glanced at the steam gage. The pressure had fallen back ten pounds, but under the circumstances, Pagel was doing all that could be expected.

As the engine crossed over on the main, George looked back. Neither of the two Diesel units were showing any exhaust. Rarely in his many years of experience did George ever feel compassion for an engine, but he found himself talking to the 2908 as if she were a faithful old horse. "Steady, old girl, don't lie down on me now."

Pagel was shifting his eyes from the track ahead and back to the steam gage. His hands were busy adjusting the stoker valves. Going by the little station at Eden, they were making close to 75. Within a mile after passing over the top of the hill the speed had increased to around 90. The track was straight and level, but the cab was swaying from side to side with the slightest dip.

*　　*　　*

Whenever problems arise in the motive power department, the traveling engineer is alerted to do the trouble shooting. Due to a chronic condition in the electrical system, the streamliners frequently arrived in Chicago too late to make connection with the *East Coast Limited*. The resulting embarrassment brought with it the wrath of the Chicago officials. Fred Cutter, the traveling engineer, was given orders to see to it that 216 arrived on schedule. However, on this particular day, there was no way to get to Green Bay in time to ride the streamliner. So he took up a vigil in the dispatcher's office at Milwaukee. Each time 216 passed a station enroute, where there was a telegrapher on duty, the "IN" and "OUT" time was transmitted over the key. By the time the report came in from Appleton Junction, it had become apparent that, instead of losing time, 216 was making up time.

Cutter sat nervously chewing on a cigar. "What engine did you use to doublehead the streamliner?" he asked.

"We had the 2908 ready and waiting for 216 when they pulled into Green Bay," answered the dispatcher.

"Who's the engineer on the head end?" continued Cutter.

"Buddy Williams."

"Oh," exclaimed Cutter, "now I'm beginning to get the picture. You've got 'Spark Plug' Williams teamed up with his favorite engine."

The telegrapher grabbed his pencil to record a message coming in on the wire. When the clicking stopped, he ripped off the sheet and gave it to the dispatcher.

"I don't believe it," commented the dispatcher, as he handed the message back to Cutter.

"Just like I figured," commented Cutter. "Spark Plug left Fond du Lac only eight minutes late; that means that he made up six minutes already. Let's see, now, he's got 65 miles to go with one stop in between and 57 minutes to make it. I say 216 is coming in on time!" As Cutter spoke, he slapped his leg to give added emphasis.

"I know Williams is fast, but you're expecting too much of an overaged steam engine," said the dispatcher. Soon the key started sounding off again. This time the telegrapher read the message aloud as he copied it down. "216 by Campbellsport at 5:32." Several dashes followed, then the report ended with . . . "going like h - - - ." The notation on the end brought a laugh from everyone in the office.

Cutter looked at the dispatcher and said, "I'll bet a box of the finest cigars that Spark Plug will bring 216 in on time."

"I'll just take that bet," returned the dispatcher.

Jumping up out of his chair, Cutter announced, "I'm going to the locker room and see to it that the Chicago crew is ready."

When the traveling engineer entered the crew's dressing quarters, he found that they hadn't even put their work clothes on.

"Get with it, men," he shouted, "216 is going to be on time!"

"You must be kidding," one of them remarked. "We were told they would be 30 minutes late."

"Is zat so!" snapped Cutter, as he spoke through clenched teeth. Then he removed his cigar and added, "Well, those guys forgot to take into account that Spark Plug Williams was doubleheading the streamliner with the 2908. I just got a message that 216 went by Campbellsport three minutes ago." The announcement raised the eyebrows of both crews and they started to change clothes in a hurry.

Cutter rushed back upstairs. Just as he entered the office, the telegrapher announced, "216 left West Bend at 5:46."

Looking over toward the dispatcher with a broad grin, Cutter remarked, "Remember when you place that order, I smoke nothing but White Owls."

"Don't count your chickens before they're hatched," returned the chief. "It's only 18 miles to Milwaukee and he's still three minutes late."

"I'm heading down to the platform so I can get the exact minute he arrives," announced Cutter.

"Guess I'll have to follow him," commented the dispatcher. "Just to keep him honest," he added with a sly wink at the telegrapher.

*　　*　　*

In a matter of seconds after stopping at West Bend, the conductor gave the highball. George opened the throttle and checked his watch. It was just 18 miles to Milwaukee and he had 15 minutes to make it. Ahead was the finest stretch of track on the whole division.

The 2908 was performing like a thoroughbred. The crack of the exhaust soon became a throaty roar. It was double track the last eight miles and most of it was down grade. The engine began to vibrate as the speed continued to increase. The right of way was completely walled off for nearly six miles coming into Milwaukee. No more crossings to worry about.

Glancing out the window, George could see the telegraph poles go by like a picket fence. Just for a fleeting moment, his memory carried him to another time when a main driver let go on Eden Hill. "Go, you son of a gun," shouted George as he gave the throttle a final adjustment.

About two miles from the station, George started to ease off on the throttle. As the engine passed the one-mile post, George hung on the whistle cord and glanced at his watch. One mile to go and he still had over two minutes left! George did a double-take as he re-checked his watch.

"That's right," he said to himself. "Now I can coast the last mile and still be on time."

The train slowed to a crawl as it came under the shed. It took a full minute to complete the last 500 feet of the run. George couldn't resist the temptation to show off just a little.

After coming to a gentle stop, Pagel jumped off the seat box with his watch in his hand and exclaimed, "Buddy, you hit it right on the nose!"

The head brakeman pulled the pin on the tank and signalled for the engine to move ahead on an adjacent track. The relief engine was ready to back in and couple on.

While waiting for the brakeman to align the switch, Fred Cutter and the Chief Dispatcher walked over to the engine. "Hey, Spark Plug," yelled Cutter.

"Hello, Freddy, what are you doing here?" returned George.

"I'd like to have a word with you."

"Well, if you promise to behave yourself, I'll let you come up here," said George.

The friendship between Cutter and George had started way back in 1913 when they took the engineer's examination together.

Cutter and the dispatcher climbed up into the cab.

"What's on your mind, Boys? asked George.

"Buddy, we came up here to congratulate you and shake your hand," said Cutter.

George pulled off his glove and stuck out his hand. "What's the occasion?" he asked.

"The fact is," continued Cutter, "I just won a box of cigars on you."

"How's that?" inquired George.

The Chief Dispatcher broke in, "When 216 left Fond du Lac, Cutter insisted that you would make up the eight minutes by the time you arrived here, so I made a bad bet."

"Frankly," continued the Chief, "I didn't believe a steam engine could make up time on the streamliner's schedule."

"While you're passing out the bouquets, don't overlook Pagel, my fireman. He did a great job keeping this baby in steam."

Both men shook Pagel's hand.

The brakeman gave the signal to head for the roundhouse. Cutter and the dispatcher climbed off. As the engine moved away, George shouted, "Hey Chief, next time, don't sell these steam engines short."

After changing clothes, George and Pagel went over to the steak house for supper. As they entered the restaurant, Bob Dilly was standing in front of the door near the cash register and George nearly bumped into him.

"Excuse me, Bob," said George as he started to go around.

"Hey! Wait a minute," said Dilly as he grabbed George's arm. The action was obviously a friendly gesture.

"What can I do for you, Bob?" asked George.

"Do you have any idea how fast you were going?" inquired Dilly.

George and Pagel both developed an amused expression for the 2908 had no speedometer.

"Oh, I don't know," replied George. "I guess you might say we moved right along," he added.

"It may interest you to know that we touched 100 heading into North Fond du Lac and the needle on my speedometer rocked between 102 and 106 for over four miles after we left West Bend."

"Not too bad for a tired old steam engine," commented George.

"Better not let Hoffman hear about it," said Pagel. "He ran himself out of gas trying to warn Buddy about going over 85."

"Well, just for the record," said Dilly, "I've been on this job since the Diesel took over and to my knowledge, that is the fastest time ever made on this run."

"You know, Bob," returned George, "our conductor deserves a lot of credit for saving us time loading and unloading."

"Speaking of the conductor," answered Dilly, "he told me to tell you that the train rode those curves beautifully. There was not a drop of coffee spilled in the diner."

"We certainly had a good crew and a wonderful engine. What else could a fella ask for?" replied George.

"If it is all the same with you boys," interrupted Pagel, "I'm going over to the counter and feed my face."

"See you around, Bob," said George, as he sat down alongside of his fireman.

*   *   *

Crude as steam engines may seem, there was a nobility about them that will never be matched by any sophisticated Diesel. In many ways they were like faithful horses, very responsive to proper care. In times of emergency they could be counted on to perform beyond their designed capacity. With a skillful hand on the throttle, a steam engine would almost seem to prance as it took off in a cloud of smoke.

Early in his career, George gained a reputation as an outstanding engineman. During the years that followed, he developed qualities of leadership. There was a sort of "Let's show the boys it can be done" way about him, such that the crews who worked with him expected performance beyond the ordinary. In the dressing rooms, cabooses and union meetings, the men exchanged many of these exciting experiences until Buddy Williams became a legend.

In 1955, after 46 years of faithful service on the Chicago & North Western Railway, Buddy Williams took his well-deserved pension. Three generations of Williamses contributed over 110 years on steam locomotives. Buddy Williams fulfilled his childhood ambition, which was to be an engineer like his dad.

At the conclusion of his final run, there was no fanfare, no officials to greet him, just warm handclasps from the men of his crew. Strangely enough, that's just the way he wanted it.

The author felt his unusual record deserved some recognition. When he expressed this view, George remarked, "No, Son, I was paid to do a job and I did it. I loved my work, and it was the many wonderful friends that made the job worthwhile." After a reflective pause, he continued, "Nearly all of my railroading pals have gone on to their reward. The days of the old iron horse are forgotten along with the men who ran them." There was a note of sadness in his voice.

Having always had the highest respect for my father, I was reluctant to disagree. But in this regard I determined to prove him wrong.

With considerable misgivings he agreed to cooperate in the writing of a book. Often, as we met for taping sessions, Dad would lose himself in his memories. During these times, his face would light up and his blue eyes just sparkled. Occasionally, he would act out some particularly dramatic scene. Sometimes he roared with laughter; a few times I saw tears flow down his cheeks as he spoke of a departed friend.

Because he was blessed with an animated personality, his vivid descriptions became indelibly etched on my memory. I endeavored to capture the characters and the experiences as faithfully as possible. Being an aircraft engineer by profession hardly qualifies one as a writer, but it was a labor of love. And if the reader has acquired a new appreciation of both the men and machines that helped make our nation great, my effort will have been repaid.

Williams, George H 1919–
    Life on a locomotive; the story of Buddy Williams,
C & NW engineer, by George H. Williams. Berkeley,
Calif., Howell-North Books [1971]

    219 p. illus., maps, ports. 24 cm. $5.95

308194   1. Williams, George H., 1890–   I. Title.   2. Chicago and
North Western Railway.
    TF140.W47W5   625.1'00924   [B]   79–165591
    ISBN 0–8310–7084–6   MARC
    Library of Congress   71 [4]